PROCESS CONTROL FOR PRACTITIONERS

PROCESS CONTROL FOR PRACTITIONERS

Contributors :
Liang Sun, Yan Li,
Aimin Li, *et al.*

AURIS REFERENCE LTD.
London, UK

Process Control for Practitioners
Contributors : Liang Sun, Yan Li, Aimin Li, *et al.*

Auris Reference Ltd., UK

www.aurisreference.com

United Kingdom

Copyright 2016

Printed in 2017 for Sale in the Indian Subcontinent

Process Control for Practitioners

ISBN: 978-1-78154-503-4

British Library Cataloguing in Publication Data
A CIP record for this book is available from the British Library

Exclusively distributed by CBS Publishers & Distributors Pvt. Ltd.

Sales & Distribution Rights only for India, Pakistan, Bangladesh, Sri Lanka, Nepal and Bhutan.This book is not to be sold outside these territories.

PREFACE

Process Control is the active changing of the process based on the results of process monitoring. Once the process monitoring tools have detected an out-of-control situation, the person responsible for the process makes a change to bring the process back into control.

It is an engineering discipline that deals with architectures, mechanisms and algorithms for maintaining the output of a specific process within a desired range. Process control is extensively used in industry and enables mass production of consistent products from continuously operated processes such as oil refining, paper manufacturing, chemicals, power plants and many others. Process control enables automation, by which a small staff of operating personnel can operate a complex process from a central control room.

Process control may either use feedback or it may be open loop. Control may also be continuous or cause a sequence of discrete events, such as a timer on a lawn sprinkler or controls on an elevator.

This page left intentionally blank.

CONTENTS

This page left intentionally blank.

LIST OF CONTRIBUTORS

Liang Sun

State Key Laboratory of Pollution Control and Resources Reuse, School of the Environment, Nanjing University, Nanjing 210023, China; E-Mails: sunliang-phd@tju.edu.cn (L.S.); liyan_0921@126.com (Y.L.)

Yan Li

State Key Laboratory of Pollution Control and Resources Reuse, School of the Environment, Nanjing University, Nanjing 210023, China; E-Mails: sunliang-phd@tju.edu.cn (L.S.); liyan_0921@126.com (Y.L.)

Aimin Li

State Key Laboratory of Pollution Control and Resources Reuse, School of the Environment, Nanjing University, Nanjing 210023, China; E-Mails: sunliang-phd@tju.edu.cn (L.S.); liyan_0921@126.com (Y.L.)

This page left intentionally blank.

Chapter 1

THE ORGANIZATIONAL CONTROL PROCESS

The control process involves carefully collecting information about a system, process, person, or group of people in order to make necessary decisions about each. Managers set up control systems that consist of four key steps:

1. **Establish standards to measure performance.** Within an organization's overall strategic plan, managers define goals for organizational departments in specific, operational terms that include standards of performance to compare with organizational activities.

2. **Measure actual performance.** Most organizations prepare formal reports of performance measurements that managers review regularly. These measurements should be related to the standards set in the first step of the control process. For example, if sales growth is a target, the organization should have a means of gathering and reporting sales data.

3. **Compare performance with the standards.** This step compares actual activities to performance standards. When managers read computer reports or walk through their plants, they identify whether actual performance meets, exceeds, or falls short of standards. Typically, performance reports simplify such comparison by placing the performance standards for the reporting period alongside the actual performance for the same period and by computing the variance — that is, the difference between each actual amount and the associated standard.

4. **Take corrective actions.** When performance deviates from standards, managers must determine what changes, if any, are necessary and how to apply them. In the productivity and quality-centered environment, workers and managers are often empowered to evaluate their own work. After the evaluator determines the cause or causes of deviation, he or she can take the fourth

step—corrective action. The most effective course may be prescribed by policies or may be best left up to employees' judgment and initiative.

These steps must be repeated periodically until the organizational goal is achieved.

CONTROL IN ORGANIZATIONS

Organizations need controls in order to determine if their goals are being met and to take corrective action if necessary.

The Nature of Control in Organizations

Control is the regulation of organizational activities so that some targeted element of performance remains within acceptable limits.

The Purpose of Control

Control provides an organization with ways to :

Adapting to Environmental Change

A control system can to anticipate, monitor, and respond to changing environmental conditions.

Limiting the Accumulation of Error

A control system will limit the number of errors that can accumulate discovery and problem solution..

Coping with Organizational Complexity

A factor increasing dramatically over recent times.

Minimizing Costs

If it is practiced effectively, control can help reduce costs and increase output.

Types of Control

Organizations establish controls in a number of different areas and at different levels. The responsibility for managing control is extensive.

Areas of Control

The four basic organizational resources usually define the areas of control.

Physical resources: Control includes inventory management, quality control and equipment control.

Human resources: Control includes selection and placement, training and development, performance appraisal and compensation.

Information resources: Control includes sales/marketing forecasting, environmental analysis, public relations, production scheduling and economic forecasting.

Financial resources: Control involves managing the organizations debt, cash flow and receivables/payables. Control of financial resources may be the most important control of all.

Control Levels

Control is practiced at many levels in the organization

o   Strategic control.

o   Structural control.

o   Operations control.

o   Financial control.

Responsibilities for Control

Ultimate responsibility for control rests with all managers throughout an organization.

Steps in the Control Process

Establishing Standards

A control standard is a target against which subsequent performance will be compared. Such standards need to be measurable and consistent with the organization's goals and should identify performance indicators.

Measuring Performance

Performance measurement is constant and ongoing in most organizations. Performance measure must be valid for control to be effective.

Comparing Performance Against Standards

Performance may be higher than, lower than, or identical to the standard.

Considering Corrective Action

After performance has been compared to standards, one of three actions is appropriate: do nothing (maintain status quo), correct the deviation, or change the standard.

OPERATIONS CONTROL

Control of the processes an organization uses to transform resources into products or services is operations control.

Preliminary Control

Preliminary control, also known as steering control or feed forward control, focuses on the resources that the organization brings in from the environment. It attempts to monitor the quality or quantity of these resources before they enter the organization.

Screening Control

Screening control, also known as yes/no control or concurrent control focuses on meeting standards for product or service quality or quantity during the transformation process. Screening control relies on feedback processes. For example, when quality checks are used to provide feedback to workers manufacturing a product, the workers know what, if any, corrective actions to take.

Post action Control

Post action control, also known as feedback control, focuses on the outputs of the organization after the transformation process is complete. Although post action control used alone may not be as effective as preliminary or screening control, it can provide management with information for future planning. Post action control also may be used as a basis for rewarding employees.

FINANCIAL CONTROL

The control of financial resources as they flow into the organization, are held by the organization, or flow out of the organization is known as financial control.

Budgetary Control

A budget is a plan expressed in numerical terms: dollars, units of output, time, or any other quantifiable factor. Budgets provide a method for measuring performance across different units within the organization. Budgets have four primary purposes: helping managers coordinate resources and projects, helping define the established control standards, providing clear guidelines about the organization's resources and expectations, and enabling organizations to evaluate the performance of managers and units.

Types of Budgets

Financial budget shows the sources and uses of cash.

Operating budget shows what quantities of products or services the organization intends to create and what financial resources will be used to create them.

Non-monetary budget expresses planned operations in non-financial terms such as units of output and machine hours.

Developing Budgets

Many organizations now allow all managers to participate in the budget process.

Strengths and Weaknesses of Budgeting

Budgets facilitate effective control and coordination and communication between departments. But budgets may be applied too rigidly; the process of developing them can be time consuming; and they may limit innovation and change.

Other Tools of Financial Control

Budgets are the most common means of financial control, but there are other useful tools: financial statements, ratio analysis, and financial audits.

Financial Statements

A profile of some aspect of an organization's financial circumstances is a financial statement. The two most commonly used financial statements are the balance sheet and the income statement. The balance sheet shows a snapshot profile of the organization's financial position. The income statement summarizes financial performance over a period of time.

Ratio Analysis

Financial ratios compare different elements of a balance sheet or income statement to one another. Ratio analysis is the calculation of one or more financial ratios to assess some aspect of the financial health of an organization. Five commonly used financial ratios are liquidity, debt, return, coverage, and operating.

Financial Audits

Audits are independent appraisals of an organization's accounting, financial, and administrative procedures. An external audit is a financial appraisal conducted by experts who are not employees of the organization. An internal audit is an appraisal conducted by employees of the organization. The objective of these audits is to verify the accuracy of financial and account procedures. Internal audits also assess these procedures for efficiency and appropriateness.

STRUCTURAL CONTROL

Structural control focuses on how well an organization's structural elements serve their intended purpose. The different approaches to structural control: bureaucratic control and clan control. Organizations characterized by these opposite approaches differ structurally in terms of foals, degree of formality, performance focus, organization design, reward system, and level of employee participation.

Bureaucratic Control

Bureaucratic control is characterized by formal and mechanistic structural arrangements. Organizations that use it tend to rely on strict rules and to have a rigid hierarchy.

Clan Control

Clan control is characterized by informal and organic structural arrangements. The goal of clan control is gaining employee commitment.

STRATEGIC CONTROL

Strategic control focuses on how effectively the organization's strategies result in the attainment of goals. The assessment of strategy requires the organization to integrate strategy and control.

Integrating Strategy and Control

Strategic control generally concentrates on organization structure, leadership, technology, human resources, and information and operations control systems. They often are seen as areas in which a strategy is or is not being effectively implemented. Strategic control focuses on the extent to which implemented strategy achieves the organization's strategic goals. If goals are not being attained, the firm will find it necessary to make changes in one or more areas,

MANAGING CONTROL IN ORGANISATIONS

Effective control successfully regulates and monitors organizational activities. Managers must be able to recognize effective control and understand how to overcome occasional resistance to control.

Characteristics of Effective Control

Effective control systems are closely integrated with planning and are flexible, accurate, timely, and objective.

Integration with Planning

The more explicit and precise the linkage between planning and control is, the more effective the control system will be. The best way to integrate control with planning is to prepare control standards as plans are being made.

Flexibility

A control system must be able to accommodate change. Unforeseen situations can create havoc with control systems that are not flexible.

Accuracy

Inaccurate information can lead to inappropriate managerial action. A control system should provide accurate information.

Timeliness

A control system should provide information as often as necessary. Generally, the more uncertain and unstable the environment, the more frequently will information be needed.

Objectivity

A control system should provide information that is as objective as possible. But even when the information is objective, managers need to look beyond the numbers when assessing performance.

Resistance to Control

In spite of the benefits that control provides, some employees resist control, especially if they feel over controlled, if they think control is inappropriately focused or rewards inefficiency, or if they are uncomfortable with accountability.

Over Control

When organizations try to control too many things, employees may feel over controlled if the controls directly affect their behaviour. When employees think that attempts to limit their behaviour are unreasonable, trouble can occur.

Inappropriate Focus

If the control system is too narrow or focuses too much on quantifiable variables, leaving no room for analysis or interpretation, it may be judged as having an inappropriate focus. When this happens, employees resist the intent of the control system by focusing their efforts only at the performance indicators being used.

Rewards for Inefficiency

If control systems knowingly or unknowingly reward inefficient activity, employees likely will resist by behaving in ways that run counter to the organization's intent.

Too Much Accountability

Effective control lets managers determine whether employees are doing their jobs. People who do not want to be answerable for their mistakes or who do not want to work as hard as their boss might like them to work are likely to resist control.

Overcoming Resistance to Control

The best way to overcome resistance to control is to design effective controls initially. Two other ways to overcome resistance are encouraging participation and developing verification procedures.

Encouraging Employee Participation

When employees are involved with planning and implementing a control system, they are less likely to resist it.

Developing Verification Procedures

The accuracy of performance indicators can be verified by having multiple standards and information systems. These checks and balances can be important verification procedures.

CONTROL SYSTEM

A control system is a device, or set of devices, that manages, commands, directs or regulates the behaviour of other devices or systems. Industrial control systems are used in industrial production for controlling equipment or machines.

There are two common classes of control systems, open loop control systems and closed loop control systems. In open loop control systems output is generated based on inputs. In closed loop control systems current output is taken into consideration and corrections are made based on feedback. A closed loop system is also called a feedback control system. The human body is a classic example of feedback systems.

Overview

The term "control system" may be applied to the essentially manual controls that allow an operator, for example, to close and open a hydraulic press, perhaps including logic so that it cannot be moved unless safety guards are in place.

An automatic sequential control system may trigger a series of mechanical actuators in the correct sequence to perform a task. For example various electric and pneumatic transducers may fold and glue a cardboard box, fill it with product and then seal it in an automatic packaging machine. Programmable logic controllers are used in many cases such as this, but several alternative technologies exist.

In the case of linear feedback systems, a control loop, including sensors, control algorithms and actuators, is arranged in such a fashion as to try to regulate a variable at a set point or reference value. An example of this may increase the fuel supply to a furnace when a measured temperature drops. PID controllers are common and effective in cases such as this. Control systems that include some sensing of the results they are trying to achieve are making use of feedback and so can, to some extent, adapt to varying circumstances. Open-loop control systems do not make use of feedback, and run only in pre-arranged ways.

Logic Control

Logic control systems for industrial and commercial machinery were historically implemented at mains voltage using interconnected relays, designed using ladder logic. Today, most such systems are constructed with programmable logic controllers (PLCs) or microcontrollers. The notation of ladder logic is still in use as a programming idiom for PLCs.

Logic controllers may respond to switches, light sensors, pressure switches, *etc.*, and can cause the machinery to start and stop various operations. Logic systems are used to sequence mechanical operations in many applications. PLC software can be written in many different ways – ladder diagrams, SFC – sequential function charts or in language terms known as statement lists.

Examples include elevators, washing machines and other systems with interrelated stop-go operations.

Logic systems are quite easy to design, and can handle very complex operations. Some aspects of logic system design make use of Boolean logic.

On–off Control

A thermostat is a simple negative feedback controller: when the temperature (the "process variable" or PV) goes below a set point (SP), the heater is switched on. Another example could be a pressure switch on an air compressor. When the pressure (PV) drops below the threshold (SP), the pump is powered. Refrigerators and vacuum pumps contain similar mechanisms operating in reverse, but still providing negative feedback to correct errors.

Simple on–off feedback control systems like these are cheap and effective. In some cases, like the simple compressor example, they may represent a good design choice.

In most applications of on–off feedback control, some consideration needs to be given to other costs, such as wear and tear of control valves and perhaps other start-up costs when power is reapplied each time the PV drops. Therefore, practical on–off control systems are designed to include hysteresis which acts as a dead band, a region around the set point value in which no control action occurs. The width of dead band may be adjustable or programmable.

Linear Control

Linear control systems use linear negative feedback to produce a control signal mathematically based on other variables, with a view to maintain the controlled process within an acceptable operating range.

The output from a linear control system into the controlled process may be in the form of a directly variable signal, such as a valve that may be 0 or 100% open or anywhere in between. Sometimes this is not feasible and so, after calculating the current required corrective signal, a linear control system may repeatedly switch an actuator, such as a pump, motor or heater, fully on and then fully off again, regulating the duty cycle using pulse-width modulation.

Proportional Control

When controlling the temperature of an industrial furnace, it is usually better to control the opening of the fuel valve *in proportion to* the current needs of the furnace. This helps avoid thermal shocks and applies heat more effectively.

Proportional negative-feedback systems are based on the difference between the required set point (SP) and process value (PV). This difference is called the *error*. Power is applied in direct proportion to the current measured error, in the correct sense so as to tend to reduce the error and therefore avoid positive feedback. The amount of corrective action that is applied for a given error is set by the gain or sensitivity of the control system.

At low gains, only a small corrective action is applied when errors are detected. The system may be safe and stable, but may be sluggish in response to changing conditions. Errors will remain uncorrected for relatively long periods of time and the system is over-damped. If the proportional gain is increased, such systems become more responsive and errors are dealt with more quickly. There is an optimal value for the gain setting when the overall system is said to be critically damped. Increases in loop gain beyond this point lead to oscillations in the PV and such a system is under-damped.

In real systems, there are practical limits to the range of the manipulated variable (MV). For example, a heater can be off or fully on, or a valve can be closed or fully open. Adjustments to the gain simultaneously alter the range of error values over which the MV is between these limits. The width of this range, in units of the error variable and therefore of the PV, is called the *proportional band* (PB). While the gain is useful in mathematical treatments, the proportional band is often used in practical situations. They both refer to the same thing, but the PB has an inverse relationship to gain – higher gains result in narrower PBs, and *vice versa*.

Under-damped Furnace Example

In the furnace example, suppose the temperature is increasing towards a set point at which, say, 50% of the available power will be required for steady-state. At low temperatures, 100% of available power is applied. When the process value (PV) is within, say 10° of the SP the heat input begins to be reduced by the proportional controller (note that this implies a 20° proportional band (PB) from full to no power input, evenly spread around the set point value). At the set point the controller will be applying 50% power as required, but stray stored heat within the heater sub-system and in the walls of the furnace will keep the measured temperature rising beyond what is required. At 10° above SP, we reach the top of the proportional band (PB) and no power is applied, but the temperature may continue to rise even further before beginning to fall back. Eventually as the PV falls back into the PB, heat is applied again, but now the heater and the furnace walls are too cool and the temperature falls too low before its fall is arrested, so that the oscillations continue.

Over-damped Furnace Example

The temperature oscillations that an under-damped furnace control system produces are unacceptable for many reasons, including the waste of fuel and time (each oscillation cycle may take many minutes), as well as the likelihood of seriously overheating both the furnace and its contents.

Suppose that the gain of the control system is reduced drastically and it is restarted. As the temperature approaches, say 30° below SP (60° proportional band (PB)), the heat input begins to be reduced, the rate of heating of the furnace has time to slow and, as the heat is still further reduced, it eventually is brought up to set point, just as 50% power input is reached and the furnace is operating as required. There was some wasted time while the furnace crept to its final temperature using only 52% then 51% of available power, but at least no harm was done. By carefully increasing the gain (*i.e.* reducing the width of the PB) this over-damped and sluggish behaviour can be improved until the system is critically damped for this SP temperature. Doing this is known as 'tuning' the control system. A well-tuned proportional furnace temperature control system will usually be more effective than on-off control, but will still respond more slowly than the furnace could under skillful manual control.

PID Control

Apart from sluggish performance to avoid oscillations, another problem with proportional-only control is that power application is always in direct proportion to the error. In the example above we assumed that the set temperature could be maintained with 50% power. What happens if the furnace is required in a different application where a higher set temperature will require 80% power to maintain it? If the gain was finally set to a 50° PB, then 80% power will not be applied unless the furnace is 15° below set point, so for this other application the operators will have to remember always to set the set point temperature 15° higher than actually needed. This 15° figure is not completely constant either: it will depend on the surrounding ambient temperature, as well as other factors that affect heat loss from or absorption within the furnace.

To resolve these two problems, many feedback control schemes include mathematical extensions to improve performance. The most common extensions lead to proportional-integral-derivative control, or PID control.

Derivative Action

The derivative part is concerned with the rate-of-change of the error with time: If the measured variable approaches the set point rapidly, then the actuator is backed off early to allow it to coast to the required level; conversely if the measured value begins to move rapidly away from the set point, extra effort is applied — in proportion to that rapidity — to try to maintain it.

Derivative action makes a control system behave much more intelligently. On control systems like the tuning of the temperature of a furnace, or perhaps the motion-control of a heavy item like a gun or camera on a moving vehicle, the derivative action of a well-tuned PID controller can allow it to reach and maintain a set point better than most skilled human operators could.

If derivative action is over-applied, it can lead to oscillations too. An example would be a PV that increased rapidly towards SP, then halted early and seemed to "shy away" from the set point before rising towards it again.

Integral Action

The integral term magnifies the effect of long-term steady-state errors, applying ever-increasing effort until they reduce to zero. In the example of the furnace above working at various temperatures, if the heat being applied does not bring the furnace up to set point, for whatever reason, integral action increasingly *moves* the proportional band relative to the set point until the PV error is reduced to zero and the set point is achieved.

Ramp UP % Per Minute

Some controllers include the option to limit the "ramp up % per minute". This option can be very helpful in stabilizing small boilers (3 MBTUH), especially during the summer, during light loads. A utility boiler "unit may be required to change load at a rate of as much as 5% per minute (IEA Coal Online - 2, 2007)".

Other Techniques

It is possible to filter the PV or error signal. Doing so can reduce the response of the system to undesirable frequencies, to help reduce instability or oscillations. Some feedback systems will oscillate at just one frequency. By filtering out that frequency, more "stiff" feedback can be applied, making the system more responsive without shaking itself apart.

Feedback systems can be combined. In cascade control, one control loop applies control algorithms to a measured variable against a set point, but then provides a varying set point to another control loop rather than affecting process variables directly. If a system has several different measured variables to be controlled, separate control systems will be present for each of them.

Control engineering in many applications produces control systems that are more complex than PID control. Examples of such fields include fly-by-wire aircraft control systems, chemical plants, and oil refineries. Model predictive control systems are designed using specialized computer-aided-design software and empirical mathematical models of the system to be controlled.

Fuzzy Logic

Fuzzy logic is an attempt to apply the easy design of logic controllers to the control of complex continuously varying systems. Basically, a measurement in a fuzzy logic system can be partly true, that is if yes is 1 and no is 0, a fuzzy measurement can be between 0 and 1.

The rules of the system are written in natural language and translated into fuzzy logic. For example, the design for a furnace would start with: "If the temperature is too high, reduce the fuel to the furnace. If the temperature is too low, increase the fuel to the furnace."

Measurements from the real world (such as the temperature of a furnace) are converted to values between 0 and 1 by seeing where they fall on a triangle. Usually, the tip of the triangle is the maximum possible value which translates to 1.

Fuzzy logic, then, modifies Boolean logic to be arithmetical. Usually the "not" operation is "output = 1 - input," the "and" operation is "output = input.1 multiplied by input.2," and "or" is "output = 1 - ((1 - input.1) multiplied by (1 - input.2))". This reduces to Boolean arithmeticif values are restricted to 0 and 1, instead of allowed to range in the unit interval [0,1].

The last step is to "defuzzify" an output. Basically, the fuzzy calculations make a value between zero and one. That number is used to select a value on a line whose slope and height converts the fuzzy value to a real-world output number. The number then controls real machinery.

If the triangles are defined correctly and rules are right the result can be a good control system.

When a robust fuzzy design is reduced into a single, quick calculation, it begins to resemble a conventional feedback loop solution and it might appear that the fuzzy design was unnecessary. However, the fuzzy logic paradigm may provide scalability for large control systems where conventional methods become unwieldy or costly to derive.

Fuzzy electronics is an electronic technology that uses fuzzy logic instead of the two-value logic more commonly used in digital electronics.

Physical Implementations

Since modern small microprocessors are so cheap (often less than $1 US), it's very common to implement control systems, including feedback loops, with computers, often in an embedded system. The feedback controls are simulated by having the computer make periodic measurements and then calculate from this stream of measurements.

Computers emulate logic devices by making measurements of switch inputs, calculating a logic function from these measurements and then sending the results out to electronically controlled switches.

Logic systems and feedback controllers are usually implemented with programmable logic controllers which are devices available from electrical supply houses. They include a little computer and a simplified system for programming. Most often they are programmed with personal computers.

Logic controllers have also been constructed from relays, hydraulic and pneumatic devices as well as electronics using both transistors and vacuum tubes (feedback controllers can also be constructed in this manner).

FEATURE OF CONTROL SYSTEM

The main feature of control system is, there should be a clear mathematical relation between input and output of the system. When the relation between input and output of the system can be represented by a linear proportionality, the system is called linear control system. Again when the relation between input and output cannot be represented by single linear proportionality, rather the input and output are related by some non-linear relation, the system is referred as non-linear control system.

Requirement of Good Control System

- **Accuracy:** Accuracy is the measurement tolerance of the instrument and defines the limits of the errors made when the instrument is used in normal operating conditions. Accuracy can be improved by using feedback elements. To increase accuracy of any control system error detector should be present in control system.

- **Sensitivity:** The parameters of control system are always changing with change in surrounding conditions, internal disturbance or any other parameters. This change can be expressed in terms of sensitivity. Any control system should be insensitive to such parameters but sensitive to input signals only.

- **Noise:** An undesired input signal is known as noise. A good control system should be able to reduce the noise effect for better performance.

- **Stability:** It is an important characteristic of control system. For the bounded input signal, the output must be bounded and if input is zero then output must be zero then such a control system is said to be stable system.

- **Bandwidth:** An operating frequency range decides the bandwidth of control system. Bandwidth should be large as possible for frequency response of good control system.

- **Speed:** It is the time taken by control system to achieve its stable output. A good control system possesses high speed. The transient period for such system is very small.

- **Oscillation:** A small numbers of oscillation or constant oscillation of output tend to system to be stable.

Types of Control Systems

There are various **types of control system** but all of them are created to control outputs. The system used for controlling the position, velocity, acceleration, temperature, pressure, voltage and current *etc.* are examples of control systems. Let us take an example of simple temperature controller of the room, to clear the concept. Suppose there is a simple heating element, which is heated up as long as the electric power supply is switched on.

As long as the power supply switch of the heater is on the temperature of the room rises and after achieving the desired temperature of the room, the power supply is switched off. Again due to ambient temperature, the room temperature falls and then manually the heater element is switched on to achieve the desired room temperature again. In this way one can manually control the room temperature at desired level. This is an example of **manual control system**.

This system can further be improved by using timer switching arrangement of the power supply where the supply to the heating element is switched on and off in a predetermined interval to achieve desired temperature level of the room. There is another improved way of controlling the temperature of the room. Here one sensor measures the difference between actual temperature and desired temperature. If there is any difference between them, the heating element functions to reduce the difference and when the difference becomes lower than a predetermined level, the heating elements stop functioning.

Both forms of the system are **automatic control system**. In former one the input of the system is entirely independent of the output of the system. Temperature of the room (output) increases as long as the power supply switch is kept on. That means heating element produces heat as long as the power supply is kept on and final room temperature does not have any control to the input power supply of the system. This system is referred as **open loop control system**.

But in the later case, the heating elements of the system function, depending upon the difference between, actual temperature and desired temperature. This difference is called error of the system. This error signal is fed back to the system to control the input. As the input to output path and the error feedback path create a closed loop, this type of control system is referred as **closed loop control system**.

Hence, there are two main **types of control system**.

Open loop control system Closed loop control system

Open Loop Control System

A control system in which the control action is totally independent of output of the system then it is called **open loop control system**. Manual control system is also an open loop control system. The block diagram of open loop control system in which process output is totally independent of controller action.

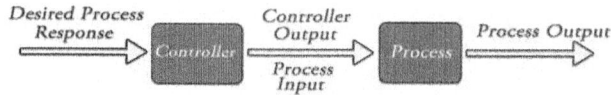

Practical Examples of Open Loop Control System :

1. **Electric Hand Drier** – Hot air (output) comes out as long as you keep your hand under the machine, irrespective of how much your hand is dried.

2. **Automatic Washing Machine** – This machine runs according to the pre-set time irrespective of washing is completed or not.

3. **Bread Toaster** - This machine runs as per adjusted time irrespective of toasting is completed or not.

4. **Automatic Tea/Coffee Maker** – These machines also function for pre adjusted time only.

5. **Timer Based Clothes Drier** – This machine dries wet clothes for pre – adjusted time, it does not matter how much the clothes are dried.

6. **Light Switch** – lamps glow whenever light switch is on irrespective of light is required or not.

7. **Volume on Stereo System** – Volume is adjusted manually irrespective of output volume level.

Advantages of Open Loop Control System :

1. Simple in construction and design.
2. Economical.
3. Easy to maintain.
4. Generally stable.
5. Convenient to use as output is difficult to measure.

Disadvantages of Open Loop Control System:

1. They are inaccurate.
2. They are unreliable.
3. Any change in output cannot be corrected automatically.

Closed Loop Control System

Control system in which the output has an effect on the input quantity in such a manner that the input quantity will adjust itself based on the output generated is called **closed loop control system**. Open loop control system can be converted in to closed loop control system by providing a feedback. This feedback automatically makes the suitable changes in the output due to external disturbance. In this way closed loop control system is called automatic control system. Figure below shows the block diagram of closed loop control system in which feedback is taken from output and fed in to input.

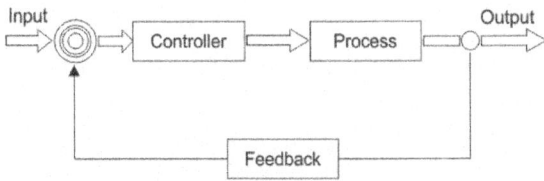

Practical Examples of Closed Loop Control System :

1. **Automatic Electric Iron** – Heating elements are controlled by output temperature of the iron.
2. **Servo Voltage Stabilizer** – Voltage controller operates depending upon output voltage of the system.
3. **Water Level Controller**– Input water is controlled by water level of the reservoir.
4. **Missile Launched & Auto Tracked by Radar** – The direction of missile is controlled by comparing the target and position of the missile.
5. **An Air Conditioner** – An air conditioner functions depending upon the temperature of the room.
6. **Cooling System in Car** – It operates depending upon the temperature which it controls.

Advantages of Closed Loop Control System :

1. Closed loop control systems are more accurate even in the presence of non-linearity.
2. Highly accurate as any error arising is corrected due to presence of feedback signal.
3. Bandwidth range is large.
4. Facilitates automation.

5. The sensitivity of system may be made small to make system more stable.

6. This system is less affected by noise.

Disadvantages of Closed Loop Control System :

1. They are costlier.

2. They are complicated to design.

3. Required more maintenance.

4. Feedback leads to oscillatory response.

5. Overall gain is reduced due to presence of feedback.

6. Stability is the major problem and more care is needed to design a stable closed loop system.

Comparison of Closed Loop And Open Loop Control System

Sr. No.	Open loop control system	Closed loop control system
1	The feedback element is absent.	The feedback element is always present.
2	An error detector is not present.	An error detector is always present.
3	It is stable one.	It may become unstable.
4	Easy to construct.	Complicated construction.
5	It is an economical.	It is costly.
6	Having small bandwidth.	Having large bandwidth.
7	It is inaccurate.	It is accurate.
8	Less maintenance.	More maintenance.
9	It is unreliable.	It is reliable.
10	Examples: Hand drier, tea maker	Examples: Servo voltage stabilizer, perspiration

Feedback Loop of Control System

A feedback is a common and powerful tool when designing a control system. Feedback loop is the tool which take the system output into consideration and enables the system to adjust its performance to meet a desired result of system.

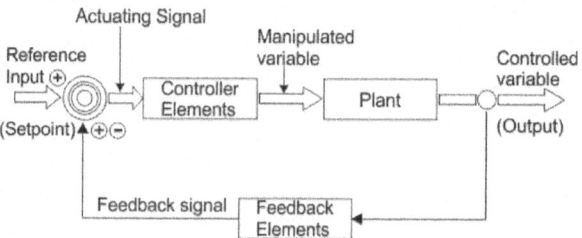

In any control system, output is affected due to change in environmental condition or any kind of disturbance. So one signal is taken from output and is fed back to the input. This signal is compared with reference input and then error signal is generated. This error signal is applied to controller and output is corrected. Such a system is called feedback system. Figure below shows the block diagram of feedback system.

When feedback signal is positive then system called positive feedback system. For positive feedback system, the error signal is the addition of reference input signal and feedback signal. When feedback signal is negative then system is called negative feedback system. For negative feedback system, the error signal is given by difference of reference input signal and feedback signal.

Effect of Feedback

Refer figure beside, which represents feedback system where R = Input signal E = Error signal G = forward path gain H = Feedback C = Output signal B = Feedback signal

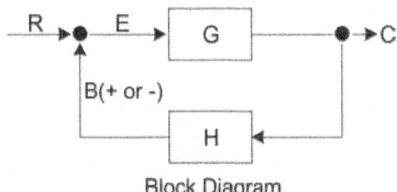

Block Diagram

1. Error between system input and system output is reduced.
2. System gain is reduced by a factor $1/(1\pm GH)$.
3. Improvement in sensitivity.
4. Stability may be affected.
5. Improve the speed of response.

Chapter 2

CLOSED-LOOP SYSTEMS

INTRODUCTION

One way in which we can accurately Control the Process is by monitoring its output and "feeding" some of it back to compare the actual output with the desired output so as to reduce the error and if disturbed, bring the output of the system back to the original or desired response. The measure of the output is called the "feedback signal" and the type of control system which uses feedback signals to control itself is called a **Close-loop System**.

A **Closed-loop Control System**, also known as a *feedback control system* is a control system which uses the concept of an open loop system as its forward path but has one or more feedback loops (hence its name) or paths between its output and its input. The reference to "feedback", simply means that some portion of the output is returned "back" to the input to form part of the systems excitation.

Closed-loop systems are designed to automatically achieve and maintain the desired output condition by comparing it with the actual condition. It does this by generating an error signal which is the difference between the output and the reference input. In other words, a "closed-loop system" is a fully automatic control system in which its control action being dependent on the output in some way.

So for example, consider our electric clothes dryer from the previous **open-loop**. Suppose we used a sensor or transducer (input device) to continually monitor the temperature or dryness of the clothes and feed a signal relating to the dryness back to the controller.

Closed-loop Control

This sensor would monitor the actual dryness of the clothes and compare it with (or subtract it from) the input reference. The error signal (error = required dryness – actual dryness) is amplified by the controller, and the controller output makes the necessary correction to the heating system to reduce any error. For example if the clothes are too wet the controller may increase the temperature or drying time. Likewise, if the clothes are nearly dry it may reduce the temperature or stop the process so as not to overheat or burn the clothes, *etc.*

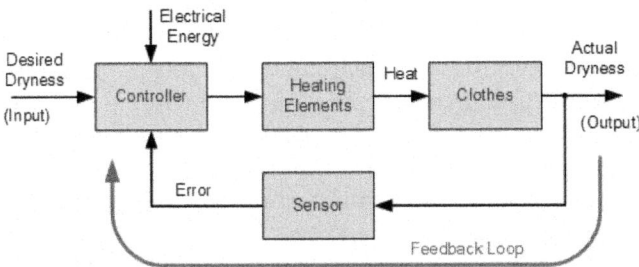

Then the closed-loop configuration is characterised by the feedback signal, derived from the sensor in our clothes drying system. The magnitude and polarity of the resulting error signal, would be directly related to the difference between the required dryness and actual dryness of the clothes.

Also, because a closed-loop system has some knowledge of the output condition, (via the sensor) it is better equipped to handle any system disturbances or changes in the conditions which may reduce its ability to complete the desired task.

For example, as before, the dryer door opens and heat is lost. This time the deviation in temperature is detected by the feedback sensor and the controller self-corrects the error to maintain a constant temperature within the limits of the preset value. Or possibly stops the process and activates an alarm to inform the operator.

A closed-loop control system the error signal, which is the difference between the input signal and the feedback signal (which may be the output signal itself or a function of the output signal), is fed to the controller so as to reduce the systems error and bring the output of the system back to a desired value. In our case the dryness of the clothes. Clearly, when the error is zero the clothes are dry.

The term **Closed-loop control** always implies the use of a feedback control action in order to reduce any errors within the system, and its "feedback" which distinguishes the main differences between an open-loop and a closed-loop system.

The accuracy of the output thus depends on the feedback path, which in general can be made very accurate and within electronic control systems and circuits, feedback control is more commonly used than open-loop or feed forward control.

Closed-loop systems have many advantages over open-loop systems. The primary advantage of a closed-loop feedback control system is its ability to reduce a system's sensitivity to external disturbances, for example opening of the dryer door, giving the system a more robust control as any changes in the feedback signal will result in compensation by the controller.

Then we can define the main characteristics of **Closed-loop Control** as being:

- To reduce errors by automatically adjusting the systems input.

- To improve stability of an unstable system.
- To increase or reduce the systems sensitivity.
- To enhance robustness against external disturbances to the process.
- To produce a reliable and repeatable performance.

Whilst a good closed-loop system can have many advantages over an open-loop control system, its main disadvantage is that in order to provide the required amount of control, a closed-loop system must be more complex by having one or more feedback paths. Also, if the gain of the controller is too sensitive to changes in its input commands or signals it can become unstable and start to oscillate as the controller tries to over-correct itself, and eventually something would break. So we need to "tell" the system how we want it to behave within some pre-defined limits.

Closed-loop Summing Points

For a closed-loop feedback system to regulate any control signal, it must first determine the error between the actual output and the desired output. This is achieved using a summing point, also referred to as a comparison element, between the feedback loop and the systems input. These summing points compare a systems set point to the actual value and produce a positive or negative error signal which the controller responds too where: Error = Set point − Actual.

The symbol used to represent a summing point in closed-loop systems block-diagram is that of a circle with two crossed lines as shown. The summing point can either add signals together in which a Plus (+) symbol is used showing the device to be a "summer" (used for positive feedback), or it can subtract signals from each other in which case a Minus (−) symbol is used showing that the device is a "comparator" (used for negative feedback) as shown.

Summing Point Types

Note that summing points can have more than one signal as inputs either adding or subtracting but only one output which is the algebraic sum of the inputs. Also the arrows indicate the direction of the signals. Summing points can be cascaded together to allow for more input variables to be summed at a given point.

Closed-loop System Transfer Function

The **Transfer Function** of any electrical or electronic control system is the mathematical relationship between the systems input and its output, and hence describes the behaviour of the system. Note also that the ratio of the output of a particular device to its input represents its gain. Then we can correctly say that the output is always the transfer function of the system times the input. Consider the closed-loop system below.

TYPICAL CLOSED-LOOP SYSTEM REPRESENTATION

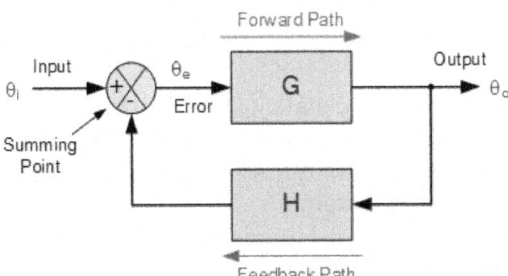

Where: block G represents the open-loop gains of the controller or system and is the forward path, and block H represents the gain of the sensor, transducer or measurement system in the feedback path.

To find the transfer function of the closed-loop system above, we must first calculate the output signal θ_o in terms of the input signal θ_i. To do so, we can easily write the equations of the given block-diagram as follows.

The output from the system is equal to:

Output = G x Error

Note that the error signal, θ_e is also the input to the feed-forward block: G

The output from the summing point is equal to:

Error = Input - H x Output

If H = 1 (unity feedback) then:

The output from the summing point will be:

Error (θ_e) = Input - Output

Eliminating the error term, then:

The output is equal to:

Output = G x (Input - H x Output)

Therefore:

G x Input = Output + G x H x Output

Rearranging the above gives us the closed-loop transfer function of:

$$\frac{\text{Output}}{\text{Input}} = \frac{\theta_o}{\theta_i} = \frac{G}{1 + GH}$$

The above equation for the transfer function of a closed-loop system shows a Plus (+) sign in the denominator representing negative feedback. With a positive feedback system, the denominator will have a Minus (−) sign and the equation becomes: 1 - GH.

We can see that when $H = 1$ (unity feedback) and G is very large, the transfer function approaches unity as:

$$\frac{\text{Output}}{\text{Input}} \rightarrow 1$$

Also, as the systems steady state gain G decreases, the expression of: $G/(1 + G)$ decreases much more slowly. In other words, the system is fairly insensitive to variations in the systems gain represented by G, and which is one of the main advantages of a closed-loop system.

Multi-loop Closed-loop System

Whilst our example above is of a single input, single output closed-loop system, the basic transfer function still applies to more complex multi-loop systems. Most practical feedback circuits have some form of multiple loop control, and for a multi-loop configuration the transfer function between a controlled and a manipulated variable depends on whether the other feedback control loops are open or closed.

Consider the multi-loop system below.

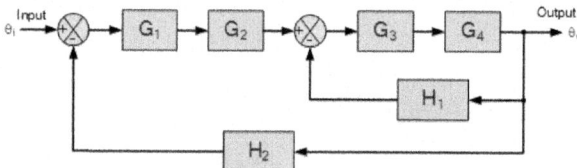

Any cascaded blocks such as G_1 and G_2 can be reduced, as well as the transfer function of the inner loop as shown.

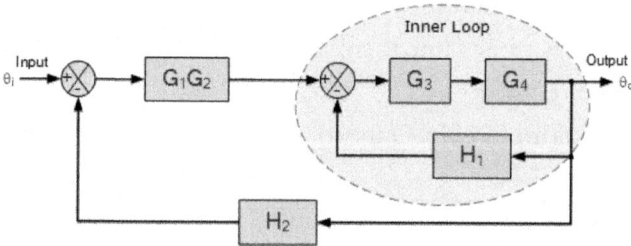

After further reduction of the blocks we end up with a final block diagram which resembles that of the previous single-loop closed-loop system.

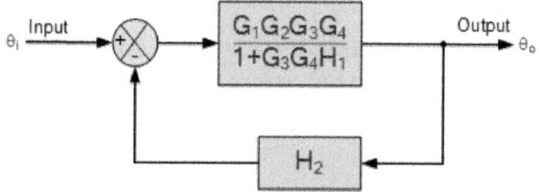

And the transfer function of this multi-loop system becomes:

$$\frac{\text{Output}}{\text{Input}} = \frac{\theta_0}{\theta_i} = \frac{G_1 G_2 G_3 G_4}{1 + G_3 G_4 H_1 + G_1 G_2 G_3 G_4 H_2}$$

Then we can see that even complex multi-block or multi-loop block diagrams can be reduced to give one single block diagram with one common system transfer function.

Closed-loop Motor Control

So how can we use **Closed-loop Systems** in Electronics. Well consider our DC motor controller from the previous open-loop. If we connected a speed measuring transducer, such as a tachometer to the shaft of the DC motor, we could detect its speed and send a signal proportional to the motor speed back to the amplifier. A tachometer, also known as a tacho-generator is simply a permanent-magnet DC generator which gives a DC output voltage proportional to the speed of the motor.

Then the position of the potentiometers slider represents the input, θ_i which is amplified by the amplifier (controller) to drive the DC motor at a set speed N representing the output, θ_0 of the system, and the tachometer T would be the closed-loop back to the controller. The difference between the input voltage setting and the feedback voltage level gives the error signal as shown.

CLOSED-LOOP MOTOR CONTROL

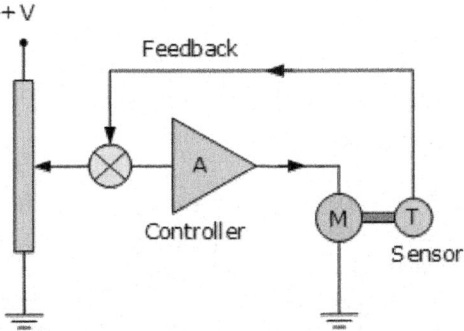

Any external disturbances to the closed-loop motor control system such as the motors load increasing would create a difference in the actual motor speed and the potentiometer input set point.

This difference would produce an error signal which the controller would automatically respond too adjusting the motors speed. Then the controller works to minimize the error signal, with zero error indicating actual speed which equals set point.

Electronically, we could implement such a simple closed-loop tachometer-feedback motor control circuit using an operational amplifier (op-amp) for the controller as shown.

Closed-loop Motor Controller Circuit

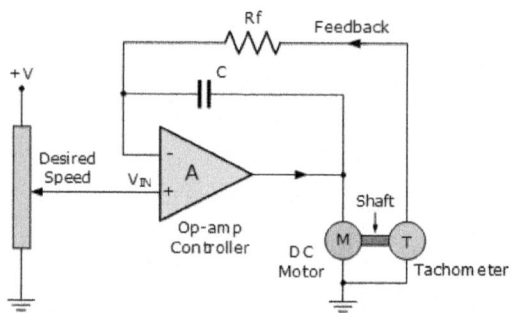

This simple closed-loop motor controller can be represented as a block diagram as shown.

Block Diagram for the Feedback Controller

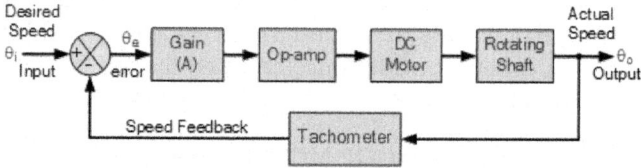

A closed-loop motor controller is a common means of maintaining a desired motor speed under varying load conditions by changing the average voltage applied to the input from the controller. The tachometer could be replaced by an optical encoder or Hall-effect type positional or rotary sensor.

Closed-loop Systems Summary

We have seen that an electronic control system with one or more feedback paths is called a **Closed-loop System**. Closed-loop control systems are also called "feedback control systems" are very common in process control and electronic control systems. Feedback systems have part of their output signal "fed back" to the input for comparison with the desired set point condition. The type of feedback signal can result either in positive feedback or negative feedback.

In a closed-loop system, a controller is used to compare the output of a system with the required condition and convert the error into a control action designed to reduce the error and bring the output of the system back to the desired response. Then closed-loop control systems use feedback to determine the actual input to the system and can have more than one feedback loop.

Closed-loop control systems have many advantages over open-loop systems. One advantage is the fact that the use of feedback makes the system response relatively insensitive to external disturbances and internal variations in system parameters such as temperature. It is thus possible to use relatively inaccurate and inexpensive components to obtain the accurate control of a given process or plant.

However, system stability can be a major problem especially in badly designed closed-loop systems as they may try to over-correct any errors which could cause the system to loss control and oscillate.

THE COMPONENTS OF A CONTROL LOOP

Components of a Control Loop

A controller seeks to maintain the measured process variable (PV) at set point (SP) in spite of unmeasured disturbances (D). The major components of a control system include a sensor, a controller and a final control element. To design and implement a controller, we must:

1) have identified a process variable we seek to regulate, be able to measure it (or something directly related to it) with a sensor, and be able to transmit that measurement as an electrical signal back to our controller, and

2) have a final control element (FCE) that can receive the controller output (CO) signal, react in some fashion to impact the process (*e.g.*, a valve moves), and as a result cause the process variable to respond in a consistent and predictable fashion.

Home Temperature Control

The home heating control system **described in this article** can be organized as a traditional control loop block diagram. Block diagrams help us visualize the components of a loop and see how the pieces are connected.

A home heating system is simple on/off control with many of the components contained in a small box mounted on our wall. Nevertheless, we introduce the idea of control loop diagrams by presenting a home heating system in the same way we would a more sophisticated commercial control application.

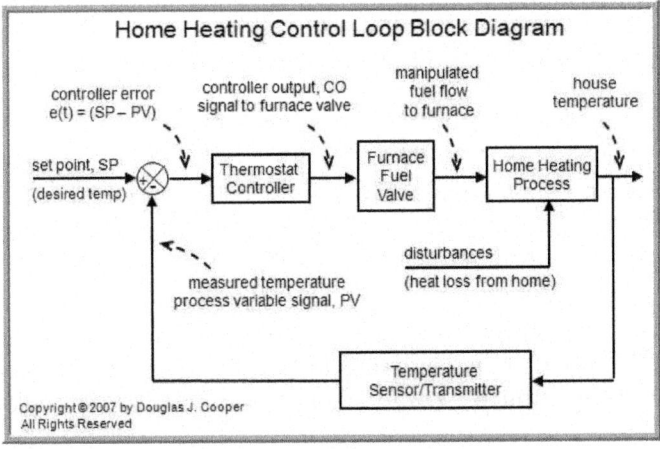

Starting from the far right in the diagram above, our process variable of interest is house temperature. A sensor, such as a **thermistor** in a modern digital thermostat, measures temperature and transmits a signal to the controller.

The measured temperature PV signal is subtracted from set point to compute controller error, e(t) = SP – PV. The action of the controller is based on this error, e(t).

In our **home heating system,** the controller output (CO) signal is limited to open/close for the fuel flow **solenoid valve** (our FCE). So in this example,

If e(t) = SP – PV > 0,

the controller signals to open the valve.

If e(t) = SP – PV < 0,

it signals to close the valve. As an aside, note that there also must be a safety interlock to ensure that the furnace burner switches on and off as the fuel flow valve opens and closes.

As the energy output of the furnace rises or falls, the temperature of our house increases or decreases and a feedback loop is complete. The important elements of a home heating control system can be organized like any commercial application:

- Control Objective: maintain house temperature at SP in spite of disturbances
- Process Variable: house temperature
- Measurement Sensor: thermistor; or bimetallic strip coil on analog models
- Measured Process Variable (PV) Signal: signal transmitted from the thermistor
- Set Point (SP): desired house temperature
- Controller Output (CO): signal to fuel valve actuator and furnace burner
- Final Control Element (FCE): solenoid valve for fuel flow to furnace
- Manipulated Variable: fuel flow rate to furnace
- Disturbances (D): heat loss from doors, walls and windows; changing outdoor temperature; sunrise and sunset; rain…

A General Control Loop and Intermediate Value Control

The home heating control loop above can be generalized into a block diagram pertinent to all feedback control loops as shown below.

Both diagrams above show a closed loop system based on negative feedback. That is, the controller takes actions that counteract or oppose any drift in the measured PV signal from set point.

While the home heating system is on/off, our focus going forward shifts to intermediate value control loops. An intermediate value controller can generate a full range of CO signals anywhere between full on/off or open/closed. The **PI algorithm** and **PID algorithm** are examples of popular intermediate value controllers.

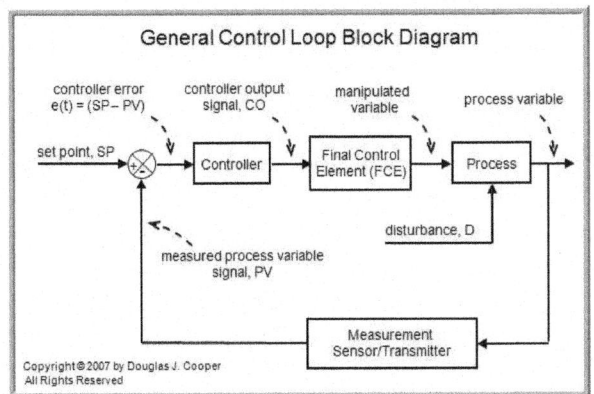

To implement intermediate value control, we require a sensor that can measure a full range of our process variable, and a final control element that can receive and assume a full range of intermediate positions between full on/off or open/closed. This might include, for example, a process valve, variable speed pump or compressor, or heating or cooling element.

Note from the loop diagram that the process variable becomes our official PV only after it has been measured by a sensor and transmitted as an electrical signal to the controller. In industrial applications. these are most often implemented as 4-20 milliamps signals, though commercial instruments are available that have been calibrated in a host of amperage and voltage units.

With the loop closed as shown in the diagrams, we are said to be in automatic mode and the controller is making all adjustments to the FCE. If we were to open the loop and switch to manual mode, then we would be able to issue CO commands through buttons or a keyboard directly to the FCE. Hence:

- open loop = manual mode
- closed loop = automatic mode

Cruise Control and Measuring Our PV

Cruise control in a car is a reasonably common intermediate value control system. For those who are unfamiliar with cruise control, here is how it works.

We first enable the control system with a button on the car instrument panel. Once on the open road and at our desired cruising speed, we press a second button that switches the controller from manual mode (where car speed is adjusted by our foot) to automatic mode (where car speed is adjusted by the controller).

The speed of the car at the moment we close the loop and switch from manual to automatic becomes the set point. The controller then continually computes and transmits corrective actions to the gas pedal (throttle) to maintain measured speed at set point.

It is often cheaper and easier to measure and control a variable directly related to the process variable of interest. This idea is central to control system design and maintenance. And this is why the loop diagrams above distinguish between our "process variable" and our "measured PV signal."

Cruise control serves to illustrate this idea. Actual car speed is challenging to measure. But transmission rotational speed can be measured reliably and inexpensively. The transmission connects the engine to the wheels, so as it spins faster or slower, the car speed directly increases or decreases.

Thus, we attach a small magnet to the rotating output shaft of the car transmission and a magnetic field detector (loops of wire and a simple circuit) to the body of the car above the magnet. With each rotation, the magnet passes by the detector and the event is registered by the circuitry as a "click." As the drive shaft spins faster or slower, the click rate and car speed increase or decrease proportionally.

So a cruise control system really adjusts fuel flow rate to maintain click rate at the set point value. With this knowledge, we can organize cruise control into the essential design elements:

- Control Objective: maintain car speed at SP in spite of disturbances
- Process Variable: car speed
- Measurement Sensor: magnet and coil to clock drive shaft rotation
- Measured Process Variable (PV) Signal: "click rate" signal from the magnet and coil
- Set Point (SP): desired car speed, recast in the controller as a desired click rate
- Controller Output (CO): signal to actuator that adjusts gas pedal (throttle)
- Final Control Element (FCE): gas pedal position
- Manipulated Variable: fuel flow rate
- Disturbances (D): hills, wind, curves, passing trucks…

The traditional block diagram for cruise control is thus:

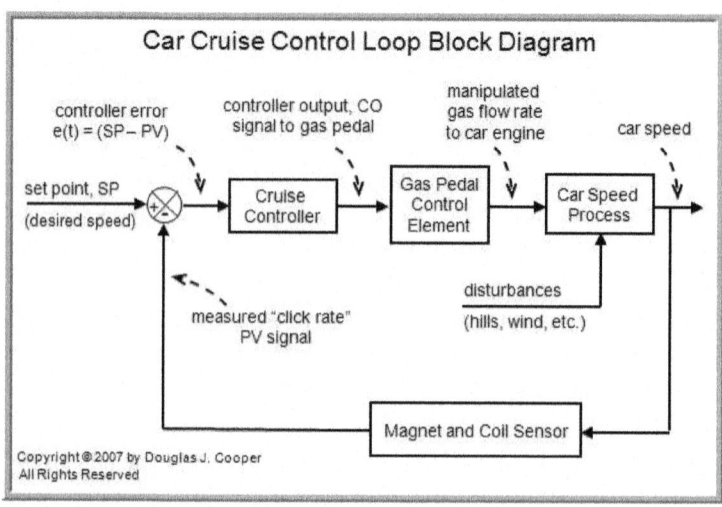

Instruments Should be Fast, Cheap and Easy

The above magnet and coil "click rate = car speed" example introduces the idea that when purchasing an instrument for process control, there are wider considerations that can make a loop faster, easier and cheaper to implement and maintain. Here is a "best practice" checklist to use when considering an instrument purchase:

- Low cost
- Easy to install and wire
- Compatible with existing instrument interface
- Low maintenance
- Rugged and robust
- Reliable and long lasting
- Sufficiently accurate and precise
- Fast to respond (small time constant and dead time)
- Consistent with similar instrumentation already in the plant

CLOSED LOOP CONTROL SYSTEMS

All automatic control systems use -ve feedback for controlling a physical parameter like position, velocity, torque *etc.* The parameter which has to be controlled is sensed by a suitable transducers and fed back to the input, for comparison with the reference value. This subtraction of the sampled output signal with that of reference input is called as –ve feedback. The difference signal, called the "error" is then amplified to drive the system (referred to as actuation) in such a manner that the output approaches the set reference value. In other words the system is designed to minimize the error signal.

All practical loads have inertia and spring constants due to which there is a delay in actuation. Hence, even though a system may be designed for –ve feedback, due to inherent time lags, the feedback may turn into +ve feedback at certain frequencies. If the loop gain is more than unity at some frequency at which the feedback is +ve, the system will oscillate. Hence, in designing control systems great care has to be taken to avoid such situations.

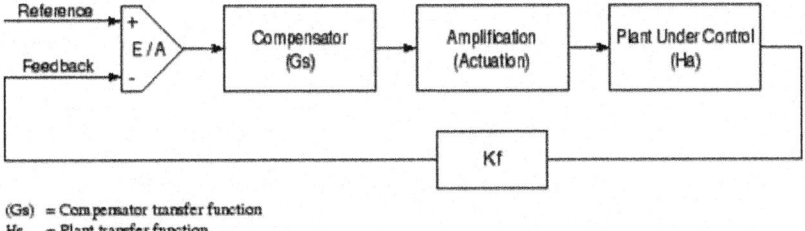

(Gs) = Compensator transfer function
Hs = Plant transfer function

Fig. : Closed loop control system.

Principles of Position Control

For controlling a heavy load, one could, use three nested feedback loops viz. a position loop, a velocity loop and a current loop. This configuration allows independent tunning of the loop parameters without affecting the adjacent loop. A current amplifier is used to amplify the current for driving the motor. The position is sensed by a suitable transducer. The velocity of the antenna is generally sensed by the tachometers mounted on the motor shaft.

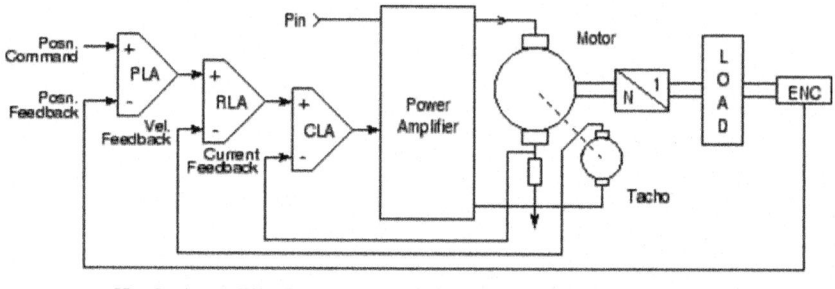

GB = Gear box.   ENC = Encoder.   PLA = Position Loop Amplifier.   RLA = Rate Loop Amplifier.

Fig. : Three nested feed back loop.

The block diagram shown above can not be directly used in all position control applications. The back-lash which is inherent in any gear box, introduces a non-linearity in the position loop. Such a system exhibits a phenomena called as "limit cycle hunting". This affects the positioning accuracy of the antenna.

Position Loop Amplifier

The position loop amplifier (PLA) has two inputs viz. command input and feedback input. In an automatic position control system, the output of the position sensor is filtered, scaled and then applied to the PLA. The command signal is applied to the other input of the PLA. The PLA (which can be either analog or digital) subtracts its two inputs to generate an error signal. This error signal is then applied to the compensator.

A compensator is designed depending on the application. For example the GMRT antennas are used for tracking of stellar radio sources which are moving at constant speed in the sky (15°/hr, the speed of the earth's rotation). For such an application, a position system having type II response is required. With a type I position compensator and with the use of rate loop in the position control, the overall system response is of type II.

Type of position system	Pointing Error	Tracking Error
Type O	Finite	Finite
Type I	Zero	Finite
Type II	Zero	Zero

Parameters like the structural natural resonant frequency (Wc) and the frictional (Bc) constants of the structure are required for the design of the position loop compensator. The main objective while designing the position compensator is that it should offer enough attenuation at the natural resonant frequency of the structure.

The output of the PLA acts as velocity command. If the target's angular position is far removed from the current position, then the error is very large and could saturate the PLA. The saturation of the PLA is considered as a fixed velocity command to the rate loop. The rate loop moves the antenna with a constant velocity towards the target position. As the antenna approaches the target position, the error at the output of the PLA goes on reducing, which commands the rate loop to reduce the speed of the antenna. When the antenna is at the target position the error at the output of the PLA goes to zero, which translates to a zero speed command to the rate loop. The sign of the error signal at the output of the PLA decides whether the antenna is to be moved forward or reverse.

Rate Loop Amplifier

The function of the Rate Loop Amplifier (RLA) is to control the velocity of the antenna. In position control applications, the rate loop improves the transient response of the position loop by adding a pole in the position loop.

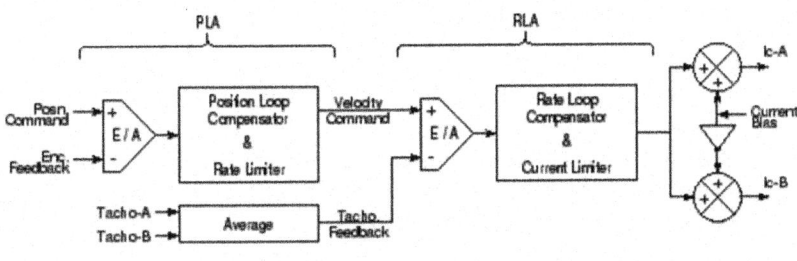

Fig. : Rate loop amplifier.

The output of the PLA which acts as a velocity command, is applied to the one input while tachometer signal is applied to the other input. The RLA subtracts both the input signals and generates an error signal which is then applied to the compensator. For position control applications like the GMRT the rate loop compensator can be of phase flag type (Type O) which avoids limit cycle hunting. The electro-mechanical time constant of the combined motor and load determines the bandwidth of the compensator. The output of the RLA acts as a command to the current loop. If the command speed is more than the actual speed, then the error at the output of the RLA becomes large, which commands the current loop to pass more current through the motor.

For GMRT antennas, where a dual drive system is used, the rate loop controls the antenna velocity by sensing the tacho signal from both the motors. Both these tacho signals are averaged and then applied to RLA as feedback. A voltage corresponding to torque bias is added/subtracted at the output of the rate loop, to generate two current commands. These two current commands are applied to the two current loop amplifiers, for controlling currents in accordance with the rate loop.

Current Loop Amplifier

The function of the Current Loop Amplifier (CLA) is to control/regulate the current of the motor which results in the control of the motor torque. The current of the motor is sensed either by a resistive shunt or with a Hall effect sensor. The control of over current should be fast in order to protect the power semiconductors during starting/stopping of the motor or in the event of fault. Also the steady state error of the current should be zero (as any error in torque affects the speed). These requirements can be met by using a "PI" (Proportional Integral) compensator.

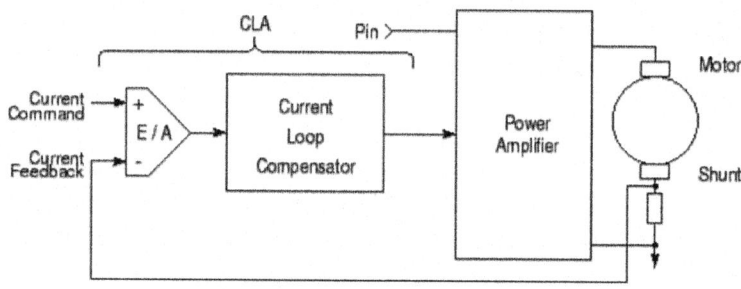

Fig. : Current loop amplifier.

The current signal is filtered, scaled and then applied to the CLA. The output of the RLA which acts as a current command, is applied to the other input. The CLA subtracts both the input signals and generates the error signal. The error signal is applied to the proportional-integral (PI) compensator. In a 3-phase SCR amplifier like one used at the GMRT, the motor current has a 150 Hz component along with the DC component. As the current is sampled and fed back to the loop amplifier, the 150 Hz component of the current gets injected into the loop. This is like injecting a noise into a system. In order to avoid oscillations in the loop, the current loop compensator is designed to heavily attenuate the 150 Hz signal component. The filtered output of the error amplifier is applied to the 4-quadrant power amplifier.

SERVO AMPLIFIERS

Servo amplifiers are 4-quadrant, regenerative power amplifiers, supplying appropriate power to the motor as commanded by a control voltage. These amplifiers are capable of suppling energy to the load, as well as absorbing energy from the load. They are designed to convert the kinetic energy of the combined motor load, into electrical energy while the load is decelerating.

The GMRT servo amplifier is a three phase, half wave, four-quadrant, fully regenerative, SCR CLA for the control of permanent magnet DC brush type motors. A CLA is a device, which keeps the current through the motor proportional to a commanded input signal.

Table: Servo amplifier specifications.

Type	3-Phase, SCR based, 4-quadrant fully regenerative.
Control Type	Phase angle control with current loop.
Input Volts	275VAC L-L, 50 Hz, 3-Phase, 4-wire.
Command Volts	+/- 10 Volt.
Maximum Current	+/- 80 Amp.
Protection	Over current & over speed.

Servo Motors

Servo motors are special category of motors, designed for applications involving position control, velocity control and torque control. These motors are special in the following ways:

1. Lower mechanical time constant.
2. Lower electrical time constant.
3. Permanent magnet of high flux density to generate the field.
4. Fail-safe electro-mechanical brakes.

For applications where the load is to be rapidly accelerated or decelerated frequently, the electrical and mechanical time constants of the motor plays an important role. The mechanical time constants in these motors are reduced by reducing the rotor inertia. Hence the rotor of these motors have an elongated structure. For DC brush type motors, the permanent magnets are mounted on the stator, while the armature conductors are on the rotor. The rotating conductors make contact with the stationary electrical source via a brush-commutator assembly. A DC tacho is mounted on the motor shaft, for indicating the shaft speed in-terms of a voltage. These motors also come with fail-safe electro-mechanical brakes. In the event of failure of the utility mains, the antennas are stopped by these brakes.

Table: Servo motor specifications.

Type	DC brush type, permanent magnet field.
Horse Power rating	6 HP.
Rated motor voltage	150 V (DC).
Rated motor current	80 Amp (Continuous).
Rated motor speed	2250 rpm.
Continuous stall torque	47 N-m.
Peak Torque	111 N-m.
Torque Sensitivity	0.56 N-m / Amp.
Back E.M.F. Constant	59 V / krpm.
Armature resistance	0.045 Ohm.
Armature inductance	0.33 mH.
Tacho sensitivity	17 V / krpm.

Gear Reducers

Generally the motors which are commercially available deliver low torque at high speed and can not be used for driving the load directly. Gear reducers are used to increase the torque so as to meet the torque demand of the load. For servo application *i.e.* for positioning the load, the gear reducers should possess following characteristics.

1. Bi-directional energy flow
2. Low back-lash
3. Low moment of inertia
4. High efficiency

The bi-directional reducers means that, the energy can be transferred from input to output as well as from output to input. During deceleration, the motor is forced to act like a generator, converting the kinetic energy of the load into electrical energy. The deceleration of the load is decided by the rate of consumption of the electrical energy produced. Planetary gear boxes meets this requirement and are hence used at the GMRT.

Position Sensors

Optical position sensors are the sensors of choice for highly accurate positioning of antennas. There are two broad styles of the encoders viz. incremental and absolute. An incremental encoder is made of a glass disc and a light interrupter. Transparent and opaque markings are put on the outer periphery of the glass disc. Light emitted from a lamp or LED is interrupted by the glass disc and received by a photo diode. As the disk rotates, the light falling on the photo detector is interrupted by the opaque markings, leading to pulses in the photdetector.

These pulses are counted to determine the change in position. The disk has an index marker, is used to provide a reference. Though incremental encoders are simple in construction and provide a cheap solution for position sensing, they suffer from one drawback. On the failure of the power to the encoder or the electronic circuit, the electronic counter looses its count value, and hence all information as to the current position. Hence, upon the resumption of the power to the antenna, one would need to move the antenna until the index marker pulse is received, a procedure called "homing". For large antennas like those at the GMRT, this is unacceptable and hence absolute encoders have to be used.

In an absolute encoder, a pattern corresponding to a gray code is printed on the glass disc. The glass disc moves through a light emitter and a set of light detectors. The number of light detectors are in proportion with the number of bits of the encoders. This enables the encoder to generate a binary word corresponding to the angular position of its shaft. The electronics housed inside the encoder converts the gray code to the natural binary. Also the parallel code gets converted into serial format for transmitting over long distance cable. The encoder is directly mounted on each axis of an antenna.

Table: Encoder specifications.

Type	Optical, absolute shaft encoder.
Resolution	17 bit (10 arcsec).
Max. Shaft speed	600 rpm.
Max. Data rate update	100 kHz.
Illumination	light emitting diode.
Input Power	+ 5V DC at 300 mA.
Output Code	Natural binary.
Output data format	Serial.
Data transmission	RS – 422 differential line driver.
Serial output	MSB first, LSB last & then parity bit.
Count Direction	CW increasing.
Operating temp.	0°C to +70°C.

Dual Drive

For a large antenna, the torque required to move the antenna is high, hence the large ratio gear reducers are used to meet the required torque demand. It is almost impossible in practice to manufacture a gear box which can deliver a large power with no back-lash. Any effort to reduce back-lash by tight coupling of pinions increases the friction of the gear box which reduces its efficiency.

With the use of large gear ratios the backlash, hysteresis, and between the motor shaft and the load shaft increases. With the increase in these parameters the nonlinearity in the position loop increases, which leads to position loop instability. There are various ways to reduce the back-lash mechanically but they are inefficient and are unsuitable for a giant antennas like those at the GMRT. Instead one uses a dual drive. Here a pair of motors, gearbox and pinion are used to drive the common load.

Two amplifiers individually drive the motors. When the load is to be held at some position, the torque produced by two motors are equal and opposite, thereby eliminating the backlash. The net torque on the load is zero hence it does not move. For a slight movement of the load in a given direction, one motor increases its torque in that direction while the other reduces its torque. The load will be subjected to a net torque which causes small movement of the load.

Digital Controller

The digital controller for GMRT antennas, is built around Intel's 8086 processor running at 8 MHz and is called as the "Station Servo Computer" The 8086 is a bus master, controlling two slave processors 8031, for analog and encoder interface. The position loop of both the axes of the GMRT servo system is implemented digitally in this servo computer. The elevation and azimuth axes angles along with time, are fed to the servo computer by the antenna base computer.

The servo computer computes the error of both the axes and performs necessary filtering (compensation). The compensator output is converted into analog signal by using 16 bit DAC and then applied to the rate loop.

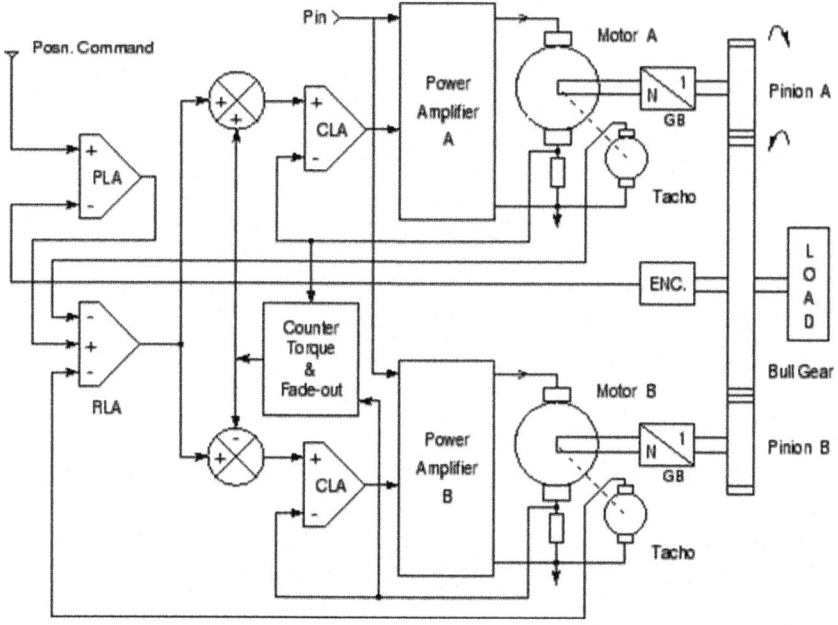

Fig. : Dual drive position control system.

For the digital implementation of a position loop, the sampling rate must be large enough. The "S" domain transfer function of the compensator is converted into a "Z" domain transfer function, by using the "Tustins approximations". The Z-domain transfer function is further converted into a difference equation, to be solve recursively at a regular interval. Tustin proposes that the sampling frequency must be greater than 10 times the compensator bandwidth. With 1.5 Hz as a structural resonant frequency of the GMRT antennas, the position loop bandwidth can be around 0.4 Hz to 0.5 Hz. For a 0.5 Hz loop bandwidth the sampling rate should be more than 5 Hz. This sets the lower limit of the sampling rate. The upper limit of the sampling rate is determined by the processor speed, other tasks of the processor, the transport lag *etc.* We have chosen 10 Hz as a sampling rate. The processor is interrupted at regular interval of 100 ms to run the real time programme.

Servo Operational Commands

The central control station sends commands to a group of antennas via an optical fiber link. Some of the operational commands, related to the servo is described next.

1. COLDSTART: On receiving this command, the servo system removes the stow-lock pins, releases the motor brakes, enables the servo amplifiers, holds both the axes at the current angle and waits for next command.

2. MV arg1,arg2: Move along the azimuth and elevation axes to the angles arg1 and arg2 respectively. The servo system releases the motor and moves the antenna.

3. TRACK arg1,arg2,arg3: Track in azimuth and elevation axes with the destination angle as arg1 and arg2 and the time parameter as arg3.

4. HOLD: Holds both the axes. On receiving this command, servo system releases brakes of both axes motors and holds the antenna in position.

5. STOP: Stops both the axes. On receiving this command, servo system disables amplifiers & applies brakes to both axes motors.

6. CLOSE: Close the observations. On receiving this command, servo system positions the elevation axis to 90:00:00 deg., disables all amplifiers, applies brakes to all motors & inserts the stow-lock pin.

7. STOW: Inserts the stow-in pin in the elevation axis and locks the axis.

8. SWRELE: Releases stow-in pin from the elevation axis and frees the axis.

9. RSTSERVO: Resets the station servo computer.

Chapter 3

FEEDBACK CONTROL AND SYSTEM

FEEDBACK CONTROL

There are many different control mechanisms that can be used, both in everyday life and in chemical engineering applications. Two broad control schemes, both of which encompass each other are *feedback control* and *feed-forward control*. *Feedback control* is a control mechanism that uses information from measurements to manipulate a variable to achieve the desired result. *Feed-forward control*, also called anticipative control, is a control mechanism that predicts the effects of measured disturbances and takes corrective action to achieve the desired result.

Feedback control is employed in a wide variety of situations in everyday life, from simple home thermostats that maintain a specified temperature, to complex devices that maintain the position of communication satellites. Feedback control also occurs in natural situations, such as the regulation of blood-sugar levels in the body. Feedback control was even used more than 2,000 years ago by the Greeks, who manufactured such systems as the float valve which regulated water level. Today, this same idea is used to control water levels in boilers and reservoirs.

Feedback Control

In feedback control, the variable being controlled is measured and compared with a target value. This difference between the actual and desired value is called the error. Feedback control manipulates an input to the system to minimize this error. An overview of a basic feedback control loop. The error in the system would be the *Output - Desired Output*. Feedback control reacts to the system and works to minimize this error. The desired output is generally entered into the system through a user interface. The output of the system is measured (by a flow meter, thermometer or similar instrument) and the difference is calculated. This difference is used to control the system inputs to reduce the error in the system.

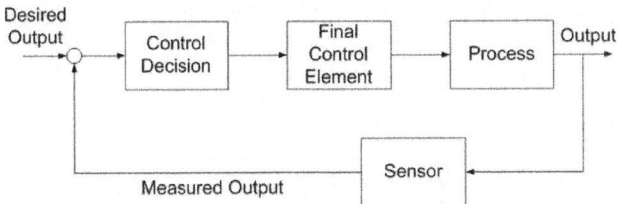

Fig. : Feedback control loop.

To understand the principle of feedback control. In order to bake cookies, one has to preheat an electric oven to 350°F. After setting the desired temperature, a sensor takes a reading inside the oven. If the oven is below the set temperature, a signal is sent to the heater to power on until the oven heats to the desired temperature. In this example, the variable to be controlled (oven temperature) is measured and determines how the input variable (heat into oven) should be manipulated to reach the desired value.

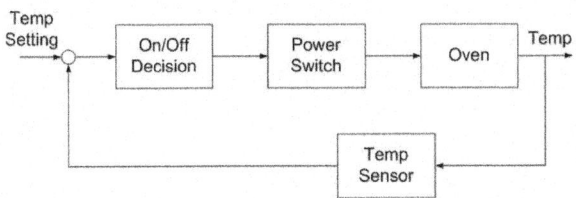

Fig. : Feedback control in an electric oven.

Feedback control can also be demonstrated with human behaviour. For example, if a person goes outside in Michigan winter, he or she will experience a temperature drop in the skin. The brain (controller) receives this signal and generates a motor action to put on a jacket. This minimizes the discrepancy between the skin temperature and the physiological set point in the person.

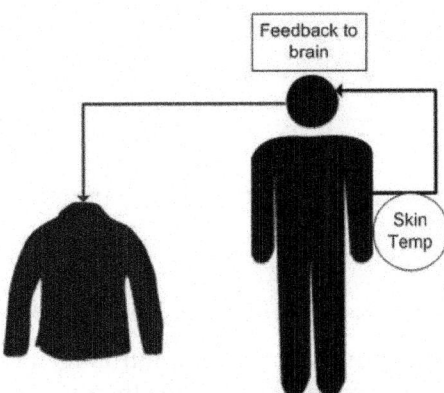

There are two types of feedback control: negative and positive. Negative feedback is the most useful control type since it typically helps a system converge toward an equilibrium state. On the other hand, positive feedback can lead a system away from an equilibrium state thus rendering it unstable, even potentially producing unexpected results. Unless stated explicitly, the term feedback control most often refers to negative feedback.

Feedback Systems

Feedback Systems process signals and as such are signal processors. The processing part of a feedback system may be electrical or electronic, ranging from a very simple to a highly complex circuits. Simple analogue feedback control circuits can be constructed using individual or discrete components, such as transistors, resistors and capacitors, etc, or by using microprocessor-based and integrated circuits (IC's) to form more complex digital feedback systems.

As we have seen, open-loop systems are just that, open ended, and no attempt is made to compensate for changes in circuit conditions or changes in load conditions due to variations in circuit parameters, such as gain and stability, temperature, supply voltage variations and/or external disturbances. But the effects of these "open-loop" variations can be eliminated or at least considerably reduced by the introduction of Feedback.

A feedback system is one in which the output signal is sampled and then fed back to the input to form an error signal that drives the system. Closed-loop Systems, we saw that in general, Feedback is comprised of a subcircuit that allows a fraction of the output signal from a system to modify the effective input signal in such a way as to produce a response that can differ substantially from the response produced in the absence of such feedback.

Feedback Systems are very useful and widely used in amplifier circuits, oscillators, process control systems as well as other types of electronic systems. But for feedback to be an effective tool it must be controlled as an uncontrolled system will either oscillate or fail to function. The basic model of a feedback system is given as:

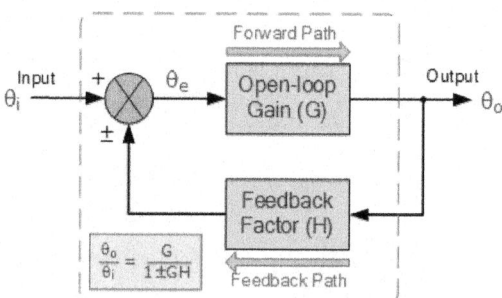

Fig. : Feedback System Block Diagram Model.

This basic feedback loop of sensing, controlling and actuation is the main concept behind a feedback control system and there are several good reasons why feedback is applied and used in electronic circuits:

- Circuit characteristics such as the systems gain and response can be precisely controlled.

- Circuit characteristics can be made independent of operating conditions such as supply voltages or temperature variations.
- Signal distortion due to the non-linear nature of the components used can be greatly reduced.
- The Frequency Response, Gain and Bandwidth of a circuit or system can be easily controlled to within tight limits.

Whilst there are many different types of control systems, there are just two main types of feedback control namely: Negative Feedback and Positive Feedback.

Positive Feedback Systems

In a "positive feedback control system", the set point and output values are added together by the controller as the feedback is "in-phase" with the input. The effect of positive (or regenerative) feedback is to "increase" the systems gain, ie, the overall gain with positive feedback applied will be greater than the gain without feedback. For example, if someone praises you or gives you positive feedback about something, you feel happy about yourself and are full of energy, you feel more positive.

However, in electronic and control systems to much praise and positive feedback can increase the systems gain far too much which would give rise to oscillatory circuit responses as it increases the magnitude of the effective input signal.

An example of a positive feedback systems could be an electronic amplifier based on an operational amplifier, or op-amp as shown.

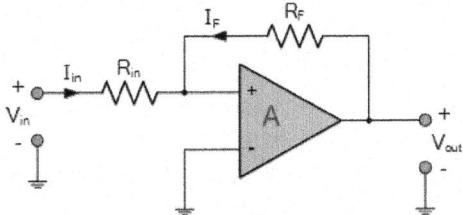

Fig. : Positive Feedback System.

Positive feedback control of the op-amp is achieved by applying a small part of the output voltage signal at V_{out} back to the non-inverting (+) input terminal via the feedback resistor, R_F.

If the input voltage V_{in} is positive, the op-amp amplifies this positive signal and the output becomes more positive. Some of this output voltage is returned back to the input by the feedback network.

Thus the input voltage becomes more positive, causing an even larger output voltage and so on. Eventually the output becomes saturated at its positive supply rail.

Likewise, if the input voltage V_{in} is negative, the reverse happens and the op-amp saturates at its negative supply rail. Then we can see that positive feedback does not allow the circuit to function as an amplifier as the output voltage quickly saturates to one supply rail or the other, because with positive feedback loops "more leads to more" and "less leads to less".

Then if the loop gain is positive for any system the transfer function will be:

$$Av = G / (1 - GH).$$

Note that if GH = 1 the system gain Av = infinity and the circuit will start to self-oscillate, after which no input signal is needed to maintain oscillations, which is useful if you want to make an oscillator.

Although often considered undesirable, this behaviour is used in electronics to obtain a very fast switching response to a condition or signal. One example of the use of positive feedback is hysteresis in which a logic device or system maintains a given state until some input crosses a preset threshold. This type of behaviour is called "bi-stability" and is often associated with logic gates and digital switching devices such as multivibrators.

We have seen that positive or regenerative feedback increases the gain and the possibility of instability in a system which may lead to self-oscillation and as such, positive feedback is widely used in oscillatory circuits such as Oscillators and Timing circuits.

Negative Feedback Systems

In a "negative feedback control system", the set point and output values are subtracted from each other as the feedback is "out-of-phase" with the original input. The effect of negative (or degenerative) feedback is to "reduce" the gain. For example, if someone criticises you or gives you negative feedback about something, you feel unhappy about yourself and therefore lack energy, you feel less positive.

Because negative feedback produces stable circuit responses, improves stability and increases the operating bandwidth of a given system, the majority of all control and feedback systems is degenerative reducing the effects of the gain.

An example of a negative feedback system is an electronic amplifier based on an operational amplifier as shown.

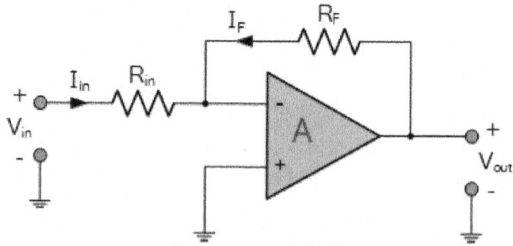

Fig. : Negative Feedback System.

Negative feedback control of the amplifier is achieved by applying a small part of the output voltage signal at V_{out} back to the inverting (–) input terminal via the feedback resistor, R_f.

If the input voltage V_{in} is positive, the op-amp amplifies this positive signal, but because its connected to the inverting input of the amplifier, and the output becomes more negative. Some of this output voltage is returned back to the input by the feedback network of R_f.

Thus the input voltage is reduced by the negative feedback signal, causing an even smaller output voltage and so on. Eventually the output will settle down and become stabilised at a value determined by the gain ratio of $R_f \div R_{in}$.

Likewise, if the input voltage V_{in} is negative, the reverse happens and the op-amps output becomes positive (inverted) which adds to the negative input signal. Then we can see that negative feedback allows the circuit to function as an amplifier, so long as the output is within the saturation limits.

So we can see that the output voltage is stabilised and controlled by the feedback, because with negative feedback loops "more leads to less" and "less leads to more".

Then if the loop gain is positive for any system the transfer function will be:

$$Av = G / (1 + GH).$$

The use of negative feedback in amplifier and process control systems is widespread because as a rule negative feedback systems are more stable than positive feedback systems, and a negative feedback system is said to be stable if it does not oscillate by itself at any frequency except for a given circuit condition.

Another advantage is that negative feedback also makes control systems more immune to random variations in component values and inputs. Of course nothing is for free, so it must be used with caution as negative feedback significantly modifies the operating characteristics of a given system.

Classification of Feedback Systems

Thus far we have seen the way in which the output signal is "fed back" to the input terminal, and for feedback systems this can be of either, Positive Feedback or Negative Feedback. But the manner in which the output signal is measured and introduced into the input circuit can be very different leading to four basic classifications of feedback.

Based on the input quantity being amplified, and on the desired output condition, the input and output variables can be modelled as either a voltage or a current. As a result, there are four basic classifications of single-loop feedback system in which the output signal is fed back to the input and these are:

- Series-Shunt Configuration – Voltage in and Voltage out or *Voltage Controlled Voltage Source* (VCVS).
- Shunt-Shunt Configuration – Current in and Voltage out or *Current Controlled Voltage Source* (CCVS).
- Series-Series Configuration – Voltage in and Current out or *Voltage Controlled Current Source* (VCCS).

- Shunt-Series Configuration – Current in and Current out or *Current Controlled Current Source* (CCCS).

These names come from the way that the feedback network connects between the input and output stages as shown.

Series-Shunt Feedback Systems

Series-Shunt Feedback, also known as *series voltage feedback*, operates as a voltage-voltage controlled feedback system. The error voltage fed back from the feedback network is in *series* with the input. The voltage which is fed back from the output being proportional to the output voltage, V_o as it is parallel, or shunt connected.

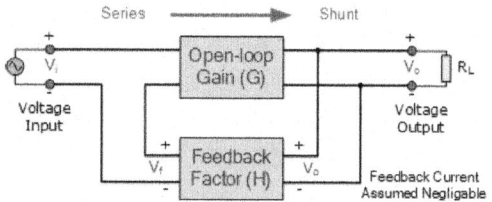

Fig. : Series-Shunt Feedback System.

For the series-shunt connection, the configuration is defined as the output voltage to the input voltage. Most inverting and non-inverting operational amplifier circuits operate with series-shunt feedback producing what is known as a "voltage amplifier". As a voltage amplifier the ideal input resistance, R_{in} is very large, and the ideal output resistance, R_{out} is very small.

Then the "series-shunt feedback configuration" works as a true voltage amplifier as the input signal is a voltage and the output signal is a voltage, so the transfer gain is given as: $Av = V_{out} \div V_{in}$.

Shunt-Series Feedback Systems

Shunt-Series Feedback, also known as *shunt current feedback*, operates as a current-current controlled feedback system. The feedback signal is proportional to the output current, Io flowing in the load. The feedback signal is fed back in parallel or *shunt* with the input as shown.

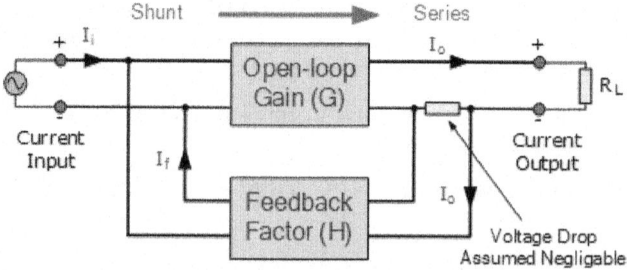

Fig. : Shunt-Series Feedback System.

For the shunt-series connection, the configuration is defined as the output current to the input current. In the shunt-series feedback configuration the signal fed back is in parallel with the input signal and as such its the currents, not the voltages that add.

This parallel shunt feedback connection will not normally affect the voltage gain of the system, since for a voltage output a voltage input is required. Also, the series connection at the output increases output resistance, R_{out} while the shunt connection at the input decreases the input resistance, R_{in}.

Then the "shunt-series feedback configuration" works as a true current amplifier as the input signal is a current and the output signal is a current, so the transfer gain is given as: $Ai = I_{out} \div I_{in}$.

Series-Series Feedback Systems

Series-Series Feedback Systems, also known as *series current feedback*, operates as a voltage-current controlled feedback system. In the series current configuration the feedback error signal is in *series* with the input and is proportional to the load current, I_{out}. Actually, this type of feedback converts the current signal into a voltage which is actually fed back and it is this voltage which is subtracted from the input.

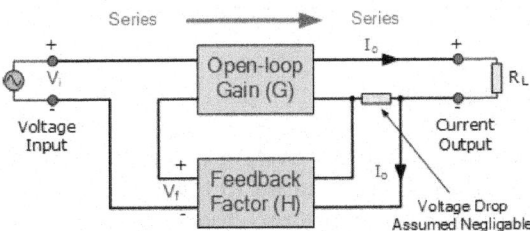

Fig. : Series-Series Feedback System.

For the series-series connection, the configuration is defined as the output current to the input voltage. Because the output current, I_o of the series connection is fed back as a voltage, this increases both the input and output impedances of the system. Therefore, the circuit works best as a transconductance amplifier with the ideal input resistance, R_{in} being very large, and the ideal output resistance, R_{out} is also very large.

Then the "series-series feedback configuration" functions as transconductance type amplifier system as the input signal is a voltage and the output signal is a current then for a series-series feedback circuit the transfer gain is given as:

$$Gm = V_{out} \div I_{in}.$$

Shunt-Shunt Feedback Systems

Shunt-Shunt Feedback Systems, also known as *shunt voltage feedback*, operates as a current-voltage controlled feedback system. In the shunt-shunt feedback configuration the signal fed back is in parallel with the input signal. The output voltage is sensed and the current is subtracted from the input current in shunt, and as such its the currents, not the voltages that subtract.

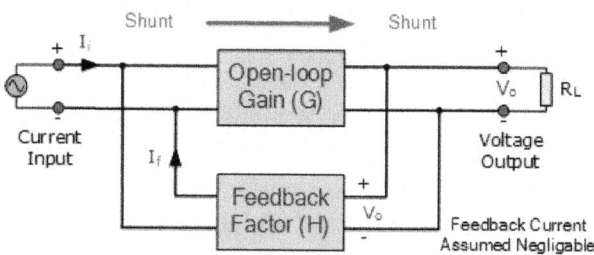

Fig. : Shunt-Shunt Feedback System.

For the shunt-shunt connection, the configuration is defined as the output voltage to the input current. As the output voltage is fed back as a current to a current-driven input port, the shunt connections at both the input and output terminals reduce the input and output impedance. therefore the system works best as a transresistance system with the ideal input resistance, R_{in} being very small, and the ideal output resistance, R_{out} also being very small.

Then the shunt voltage configuration works as transresistance type voltage amplifier as the input signal is a current and the output signal is a voltage, so the transfer gain is given as: $Rm = I_{out} \div V_{in}$.

Feedback Systems Summary

We have seen that a Feedback System is one in which the output signal is sampled and then fed back to the input to form an error signal that drives the system, and depending on the type of feedback used, the feedback signal which is mixed with the systems input signal, can be either a voltage or a current.

Feedback will always change the performance of a system and feedback arrangements can be either positive (regenerative) or negative (degenerative) type feedback systems. If the feedback loop around the system produces a loop-gain which is negative, the feedback is said to be negative or degenerative with the main effect of the negative feedback is in reducing the systems gain.

If however the gain around the loop is positive, the system is said to have positive feedback or regenerative feedback. The effect of positive feedback is to increase the gain which can cause a system to become unstable and oscillate especially if GH = -1.

We have also seen that block-diagrams can be used to demonstrate the various types of feedback systems. In the block diagrams above, the input and output variables can be modelled as either a voltage or a current and as such there are four combinations of inputs and outputs that represent the possible types of feedback, namely: Series Voltage Feedback, Shunt Voltage Feedback, Series Current Feedback and Shunt Current Feedback.

The names for these different types of feedback systems are derived from the way that the feedback network connects between the input and output stages either in parallel (shunt) or series.

TYPES OF FEEDBACK CONTROL

All the graphs shown in this chapter use parameter values for the thermal model that are typical of a small domestic cooker and the set-point temperature T_s is indicated by the red lines.

On-Off Control

This is the simplest form of control, used by almost all domestic thermostats. When the oven is cooler than the set-point temperature the heater is turned on at maximum power, M, and once the oven is hotter than the set-point temperature the heater is switched off completely. The turn-on and turn-off temperatures are deliberately made to differ by a small amount, known as the hysteresis H, to prevent noise from switching the heater rapidly and unnecessarily when the temperature is near the set-point. The fluctuations in temperature shown on the graph are significantly larger than the hysteresis, as can be confirmed with the interactive simulation, due to the significant heat capacity of the heating element.

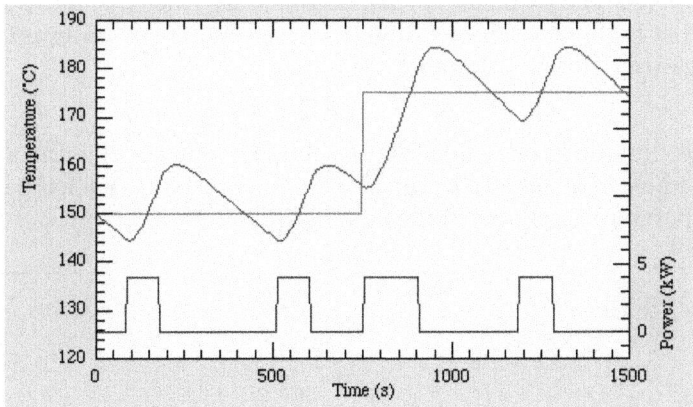

Proportional Control

A proportional controller attempts to perform better than the On-Off type by applying power, W, to the heater in proportion to the difference in temperature between the oven and the set-point,

$$W = P \times (T_s - T_o)$$

where P is known as the *proportional gain* of the controller. As its gain is increased the system responds faster to changes in set-point but becomes progressively underdamped and eventually unstable. The final oven temperature lies below the set-point for this system because some difference is required to keep the heater supplying power. The heater power must always lie between zero and the maximum M because it can only source, not sink, heat.

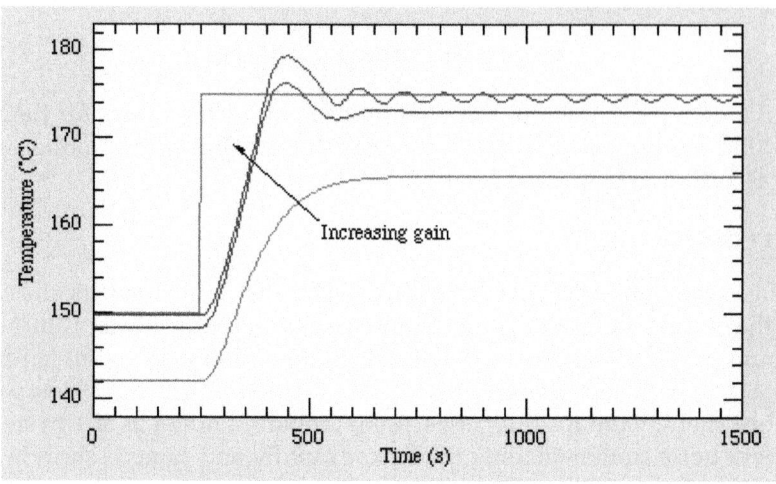

Proportional + Derivative Control

The stability and overshoot problems that arise when a proportional controller is used at high gain can be mitigated by adding a term proportional to the time-derivative of the error signal,

$$W = P \times ((T_s - T_o) + D \times \mathbf{ddt}(T_s - T_o))$$

This technique is known as *PD control*. The value of the *damping constant, D,* can be adjusted to achieve a critically damped response to changes in the setpoint temperature, as shown in the next figure.

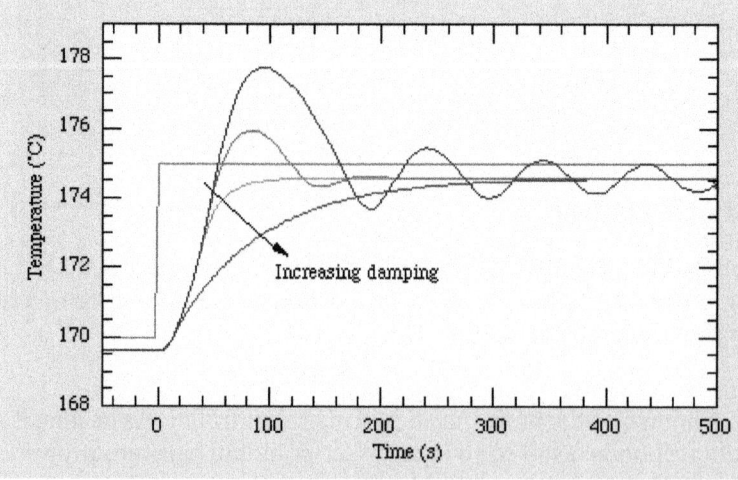

Too little damping results in overshoot and ringing, too much causes an unnecessarily slow response.

Proportional+Integral+Derivative Control

Although PD control deals neatly with the overshoot and ringing problems associated with proportional control it does not cure the problem with the steady-state error. Fortunately it is possible to eliminate this while using relatively low gain by adding an integral term to the control function which becomes

$$W = P \times ((T_s - T_o) + D \times \mathrm{ddt}(T_s - T_o) + I \times \int(T_s - T_o)\mathrm{dt})$$

where I, the *integral gain* parameter is sometimes known as the controller *reset level*. This form of function is known as proportional-integral-differential, or PID, control. The effect of the integral term is to change the heater power until the time-averaged value of the temperature error is zero. The method works quite well but complicates the mathematical analysis slightly because the system is now third-order.

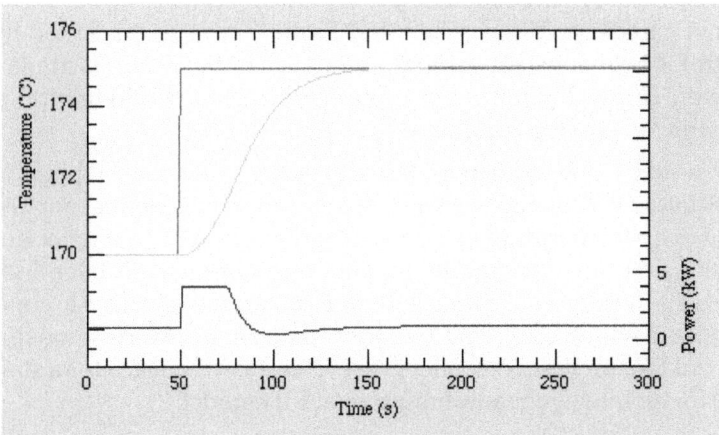

The figure shows that, as expected, adding the integral term has eliminated the steady-state error. The slight undershoot in the power suggests that there may be scope for further tweaking.

Proportional + Integral Control

Sometimes, particularly when the sensor measuring the oven temperature is susceptible to noise or other electrical interference, derivative action can cause the heater power to fluctuate wildly. In these circumstances it is often sensible use a PI controller or set the derivative action of a PID controller to zero.

Third-Order Systems

Systems controlled using an integral action controller are almost always at least third-order. Unlike second-order systems, third-order systems are fairly uncommon in physics but the methods of control theory make the analysis quite straightforward. For instance, applying the so-called *Routh-Hurwitz stability criterion*, which is a systematic way of classifying the complex roots of the auxiliary equation for the model, it can be shown that provided the integral gain is kept sufficiently small then parameter values can be found to give an acceptably damped response with the error temperature eventually tending to zero if the set-point is changed by a step or linear ramp in time. Whereas derivative control improved the system damping, integral control eliminates steady-state error at the expense of stability margin.

PRACTICAL MATTERS

In its raw form integral control can be a mixed blessing; if the error $T_s - T_o$ is large for a long period, for example after a large change in T_s or at switch-on, the value of the integral can become excessively large and cause overshoot or undershoot that takes a long time to recover. To avoid this problem, which is often called 'integral wind-up', sophisticated controllers will inhibit integral action until the system gets fairly close to equilibrium. One method of achieving this is used by the interactive simulation: when the "Limit I?" option is selected the value of the integral is held constant during periods when the heater is at maximum, or zero, power. This technique seems quite effective and would be straightforward to incorporate in a real controller.

Any system using a resistive electrical heater to control temperature is inherently non-linear because an electrical heater can only generate, not absorb, heat. When the oven temperature is higher than the set-point cooling occurs at a rate that depends on the oven and its temperature not the controller and dual PID controllers allow different heating and cooling parameter values to cope with this. It is possible to build your own PID controller from a few operational-amplifiers. Commercial PID process controllers vary in cost between £75 for a simple model and £600 for an intelligent autotuning dual PID model.

Don't just assume that the knobs on a PID controller correspond to the parameters defined in this document. Values are sometimes specified by time constants in which case a long integral time constant is equivalent to a low value of I but a long derivative time constant means a large value of D. The proportional gain is sometimes set by choosing a proportional band which is the change in input that gives maximum change in heater power so a small number for this corresponds to a large value of P.

Varieties of PID Algorithms

The *parallel algorithm* variety of PID control, the version discussed in this doument

$$W = P \times ((T_s - T_o) + D \times \mathrm{ddt}(T_s - T_o) + I \times \int (T_s - T_o)\mathrm{dt})$$

is often referred to as the 'ideal algorithm'. To implement this scheme accurately one needs at least three amplifiers (the example controller circuituses five). However, if slight deviations from the 'ideal' behaviour are permitted, only one amplifier is needed. This can be a great advantage, particularly in pneumatic systems where amplifiers are expense items. Differences in the achievable control performance due to which algorithm is being used are not normally significant. However, the tuning procedures used do differ slightly. Also, some controllers only apply derivative action to the process variable, not to the set point. Whether this is an advantage or not depends on the circumstances.

Control Theory

Avoid re-inventing the wheel when tackling difficult feedback or control problems - control theory is a well-developed branch of engineering and has a range of powerful techniques to design and analyse systems involving feedback. As well as having systematic methods for solving complicated problems it introduces the important ideas of *controllability* ('Is it possible to control X by adjusting Y?'), *observability* ('Does the system have distinct states that can't be unambiguously identified by the controller?') and *robustness* ('Will control be regained satisfactorily after an unexpected disturbance?').

Noise and the Frequency Domain

The frequency domain behaviour of the model can be investigated with the interactive simulator which will plot the open- and closed-loop frequency response for the system. As the controller gain is increased the phase margin reduces towards zero causing the overshoot described previously and a resonant peak in the frequency domain response. Any additional lag in the system, for example a non-negligible time-constant for the sensor measuring T_o, will make it possible for the system to oscillate, which is the reason for the second step in the procedure suggested for tuning a PID controller. Note that integral action reduces the phase margin, derivative action improves it. Even if a system is technically stable it is unwise to operate it with a large peak in the closed-loop gain as this will act as an amplifier for any sensor noise and may cause large and undesired fluctuations in the heater power. If you have a noisy system to control you almost certainly do not want to use any derivative action.

TUNING A PID TEMPERATURE CONTROLLER

In some case one may be able to measure the oven time constants directly and hence calculate the best controller settings. Often an equipment manufacturer will have suggested settings based on their commissioning report - a good reason read the manual first. Sometimes one has no option but to set up, or 'Tune', a system in closed-loop mode by trial and error so here are two straightforward procedures to tune a PID-controlled oven, they will get fairly close to optimum settings in most cases.

CDHW Method :

1. Adjust the set-point value, T_s, to a typical value for the envisaged use of the system and turn off the derivative and integral actions by setting their levels to zero. Select a safe value for the maximum power M and increase the proportional gain until the system is just oscillating.

2. Note the period of oscillation then reduce the gain by 30%.
3. Suddenly decreasing or increasing T_s by about 5% should induce under-damped oscillations. Try several values of derivative level and choose a value for that gives a critically damped response. If the controller is calibrated D will need to be approximately one third of the oscillation period noted above.
4. Slowly increase the integral level until oscillation just starts, then reduce this level by a factor of two or three - this should be enough to stop the oscillation. I have found it is a good idea to use the lowest integral level that gives adequate performance.
5. Check the overall performance of system is satisfactory under the conditions it will be used.

This procedure is based on the assumption that a critically damped system is optimal and the fact that stability and noise must be traded for response time. Please bear in mind that the second step may involve large temperature oscillations and so the procedure would not be suitable if these could be dangerous or cause damage, for example in a chemical processing plant.

John Shaw's (Ziegler-Nichols Based) Method :

1. Adjust the set-point value, T_s, to a typical value for the envisaged use of the system and turn off the derivative and integral actions by setting their levels to zero. Select a safe value for the maximum power M and set the proportional gain to minimum.
2. Progressively increase the gain until suddenly decreasing or increasing T_s by about 5% induces oscillations that are just self-sustaining.
3. The gain at this stage will be set to the ultimate gain G_u the period of the oscillations is known as the ultimate period t_u. Note the values of each quantity.
4. Set the controller parameters as follows:
   o P-Control: $P = 0.50*G_u$, $I = 0$, $D = 0$.
   o PI-Control: $P = 0.45*G_u$, $I = 1.2/t_u$, $D = 0$.
   o PID-Control: $P = 0.60*G_u$, $I = 2/t_u$, $D = t_u/8$.
5. Check the overall performance of system is satisfactory under the conditions it will be used.

This procedure was adapted slightly from John Shaw's, description of the Ziegler-Nichols Closed Loop method. It should yield a system that is slightly underdamped; if a less "aggressive" response is desired try reducing P to half the values listed. As was the case with the CDHW method the second step may involve large temperature oscillations and so the procedure would not be suitable if these could be dangerous or cause damage, for example in a nuclear reactor. Strictly speaking, the Ziegler-Nichols method was developed for the traditional *series*, or *interacting* design of controller.

DESIGN OF STATE-FEEDBACK CONTROL SYSTEMS

Module overview. *The main design approach for systems described in state-space form is the use of state feedback. One selects pole locations to achieve a satisfactory dynamic response and develops the control law for the closed-loop system that corresponds to satisfactory dynamic response. One has to design an estimator for the states, because these are generally not measurable. This estimator is an observer that delivers the information about the states so that they can be used for control. The combined observer-controller problem is discussed. Several pole-placement designs for controllers with proportional and integral state feedback and observers based on controller canonical form are given. A comprehensive example shows the overall design procedure.*

Module objectives. *When you have completed this module you should be able to:*

1. Design control systems using state feedback for pole placement.
2. Design observers by pole placement.
3. Transform a state-space representation into controller canonical form.

Module prerequisites. *State space representation, transfer function, poles and zeros.*

STRUCTURES AND PROPERTIES OF STATE-FEEDBACK CONTROL SYSTEMS

In the following, the design of state-feedback controllers for single-input-single-output systems described by

$$\dot{x}(t) = Ax(t) + bu(t) \qquad x(t_o) \text{ initial condition}$$

$$y(t) = c^T x(t)$$

The dynamical characteristics, for example, stability, decay of oscillations or sensitivity to disturbances, are determined by the distribution of the eigenvalues of the system matrix A in the s plane. The goal is to influence the system specifically so that it shows a desired behaviour. In the sense of command input regulation the control system is configured as shown in Figure below.

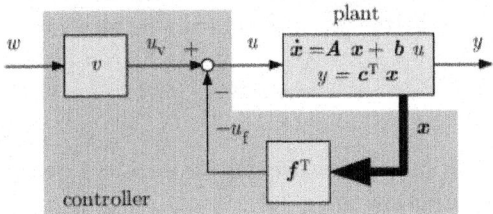

Fig. : Basic structure of a state-feedback control system.

It is assumed, that all state variables are measured. A linear combination of the state variables is fed back by

$$u_f(t) = -f^T x(t)$$

with

$$f = [f_1, \quad f_2, \quad \cdots, \quad f_n].$$

While this feedback determines the dynamical behaviour, the feedforward of the command variable w

$$u_v(t) = v\, w(t)$$

using the scalar gain v influences the static behaviour. The manipulated variable is obtained from

$$u(t) = u_f(t) + u_v(t) = -f^T x(t) + v\, w(t).$$

After the substitution of the manipulated variable into Equations, one obtains the closed-loop system as

$$\dot{x}(t) = \left[A - b\, f^T \right] x(t) + b v\, w(t) \quad x(t_o) \text{ initial condition}$$

$$y(t) = c^T x(t).$$

For the dynamical and static behaviour the following specifications must be fulfilled:

The dynamical behaviour of the closed loop should be specified by given poles. As these closed-loop poles are the eigenvalues of the closed-loop system matrix $(A - b f^T)$, the desired distribution of the eigenvalues in the s plane for this matrix is specified.

In the steady state the control error must vanish so that the plant output follows

$$y_\infty = \lim_{t \to \infty} y(t) = c^T \lim_{t \to \infty} x(t) = \lim_{t \to \infty} w(t) = w_\infty.$$

State-feedback Control in the Frequency Domain

As the dynamical behaviour of the closed loop is specified in the s domain, it is appropriate to discuss some properties in the frequency domain. The initial point of this discussion is the closed-loop transfer function

$$\frac{Y(s)}{W(s)} = G_W(s)$$

Applying the Laplace transform as shown to Equations one obtains the closed-loop transfer function as

$$G_W(s) = c^T (sI - A + b\, f^T)^{-1} b v.$$

The same procedure for the plant from Equations gives

$$G_P(s) = c^T (sI - A)^{-1} b.$$

With Equation the closed-loop and plant transfer functions are rewritten as

$$G_W(s) = c^T (\Phi^{-1}(s) + b\, f^T)^{-1} bv$$

and

$$G_P(s) = c^T \Phi b = \frac{N(s)}{D(s)}.$$

Applying the matrix inversion lemma to the inner part $(\Phi^{-1}(s) + b\, f^T)^{-1}$ of Equation, one obtains

$$G_W(s) = c^T \left\{ \Phi(s) - \Phi(s)b\left[1 + f^T\Phi(s)b\right]^{-1} f^T\Phi(s) \right\} bv$$

$$= \left\{ c^T\Phi(s)b - \frac{c^T\Phi(s)\, b\, f^T\Phi(s)b}{1 + f^T\Phi(s)b} \right\} v$$

$$= \frac{c^T\Phi(s)bv}{1 + f^T\Phi(s)b}.$$

After inserting Equations the closed-loop transfer function is given by

$$G_W(s) = \frac{c^T\Psi(s)bv}{D(s) + f^T\Psi(s)b}$$

$$= \frac{N(s)\, v}{D(s) + f^T\Psi(s)b},$$

where the denominator $\left[D(s) + f^T\Psi(s)b\right]$ is the characteristic polynomial of the closed-loop system and $D(s)$ is that of the open-loop system. As only the plant numerator polynomial $N(s)$ appears in the closed-loop transfer function, the closed-loop zeros are the same as the open-loop zeros. This means that the zeros cannot be influenced by a state-feedback controller; it only moves the poles.

Steady State of the Closed-loop System

The steady state from Equation can only be reached if the state of the closed-loop system for $t \to \infty$ approaches the final value

$$x_\infty = \lim_{t \to \infty} x(t).$$

The condition

$$\dot{x}(t) = 0 = (A - b\, f^T)\, x_\infty + bv\, w_\infty$$

is obviously valid, from which the steady state can be obtained as

$$x_\infty = -(A - b\, f^T)^{-1} bv\, w_\infty,$$

which is always possible, because for an asymptotically stable closed-loop the matrix $(A - b\,f^T)$ has always full rank. For the output, one obtains

$$y_\infty = c^T x_\infty = -c^T (A - b\,f^T)^{-1} bv\,w_\infty,$$

where it must be observed that this value must not vanish, *i.e.*

$$c^T (A - b\,f^T)^{-1} b \neq 0.$$

This condition can be fulfilled, if the closed-loop transfer function $G_W(s)$ from Eq. (13.11) has no zero at $s = 0$. From the discussion about the zeros. It is clear that the plant transfer function from Eq. (13.12) must not have a zero at $s = 0$. This means that

$$c^T A^{-1} b \neq 0.$$

With the condition given in Eq. (13.18) all given values y_∞ are reachable with command input signals that have a constant steady-state value of w_∞.

From the conditions in Eqs. (13.17) and (13.9) the feed-forward gain of the controller is obtained as

$$v = \frac{-1}{c^T (A - bf^T)^{-1} b}.$$

STATE-FEEDBACK CONTROL WITH INTEGRATOR

For a given feedback vector f the feed forward gain v can be calculated according to Equation so that the closed-loop system, the desired static behaviour with a vanishing control error. The control structure used is not very robust with respect to the control error, because this error is not fed back. Uncertainties in the plant model parameters or disturbances acting on the plant may cause steady-state control errors. In order to reject such effects, one can use a similar approach. The cascade control system has

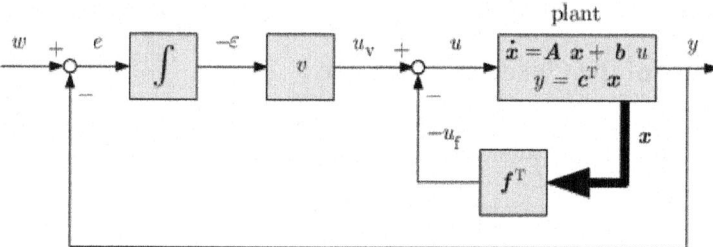

Fig. : Basic structure of a cascade state-feedback control system with integrator.

the state-feedback configuration as its inner loop. The controlled variable y is compared with the set-point value w and the control error ϵ is fed back to an integrator. The former feedforward gain v is now the gain of the integrator. From the closed-loop point of view with this configuration the gain like f is a feedback parameter. This will be shown in the following.

The equations of the closed-loop system can be directly determined as follows:

$$\dot{x}(t) = Ax(t) + bu(t) \qquad x(t_o) \text{ initial condition}$$

$$y(t) = c^T x(t)$$

$$\dot{e}(t) = y(t) - w(t)$$

$$u(t) = u_f(t) + u_v(t) = -f^T x(t) - v \in (t).$$

Combining Equation of the open-loop system in matrix notation, one obtains

$$\begin{bmatrix} \dot{x}(t) \\ \dot{e}(t) \end{bmatrix} = \begin{bmatrix} A & 0 \\ c^T & 0 \end{bmatrix} \begin{bmatrix} x(t) \\ \in (t) \end{bmatrix} + \begin{bmatrix} b \\ 0 \end{bmatrix} u(t) + \begin{bmatrix} 0 \\ -1 \end{bmatrix} w(t).$$

Equation for the manipulated variable can be represented by

$$u(t) = -\begin{bmatrix} f \\ v \end{bmatrix}^T \begin{bmatrix} x(t) \\ \in (t) \end{bmatrix}.$$

With the abbreviations

$$A^* = \begin{bmatrix} A & 0 \\ c^T & 0 \end{bmatrix}, \quad b^* = \begin{bmatrix} b \\ 0 \end{bmatrix}, \quad v^* = \begin{bmatrix} 0 \\ -1 \end{bmatrix}, x^*(t) = \begin{bmatrix} x(t) \\ \in (t) \end{bmatrix}$$

Equations are

$$\dot{x}^*(t) = A^* x^*(t) + b^* u(t) + w(t) \quad x^*(t_o) \text{ initial condition}$$

$$u(t) = -f^{*T} x^*(t)$$

and for the closed-loop system after inserting Equations

$$\dot{x}^*(t) = \begin{bmatrix} A^* - b^* f^{*T} \end{bmatrix} x^*(t) + v^* w(t) \quad x^*(t_o) \text{ initial condition}$$

which is formally the same system as that with the simple state-feedback controller in Equation. Here, the controller parameters in f^* instead of those in f must be determined for the desired specifications. This extended problem is now reduced to the original problem, and for both cases the same design procedure can be applied. It means, that for the state-feedback controller with integrator the same design procedures for the controller parameters can be applied as for the original system. This simplifies the design of the extended problem significantly.

Design of state-feedback controllers by pole placement

The difficulty of this design consists essentially in the determination of the feedback vector f so that the n eigenvalues of the system matrix $(A - bf^T)$ have the desired distribution. After that, the determination of the feed forward gain v in the control structure without integrator is very simple.

The characteristic polynomial $Q(s) = \det\begin{bmatrix} sI - (A - bf^T) \end{bmatrix}$ of the closed loop system from Equation is a monic polynomial of order n. The coefficients q_i of it are functions of the controller parameters h_i:

$$\det\left[sI-(A-b\,f^{T})\right]=q_{o}(f)+q_{1}(f)s+q_{2}(f)s^{2}+\cdots+q_{n-1}(f)s^{n-1}+s^{n}.$$

By a proper choice of the vector f this polynomial $Q(s)$ should be made equal to the desired polynomial $P(s)$ with n zeros, which are the desired poles or eigenvalues s_i, respectively, of the closed-loop system:

$$P(s)=\prod_{i=1}^{n}(s-s_{i}).$$

Multiplying all factors on the right-hand side of Equation one obtains this polynomial as

$$P(s)=p_{o}+p_{1}s+p_{1}s+p_{2}s^{2}+\cdots+p_{n-1}s^{n-1}+s^{n}.$$

For all values of s, the condition $Q(s) = P(s)$ must be fulfilled. A comparison of the corresponding terms of both sides gives the coefficients as

$$q_{0}(f)=p_{0}, \quad q_{1}(f)=p_{1}, \quad q_{2}(f)=p_{2}, \quad \cdots \quad q_{n-1}(f)=p_{n-1},$$

from which the controller parameters f_i can be obtained. This approach, however, has the following drawbacks:

The Equation for determining the controller parameters are complicated, in general nonlinear.

For higher-order systems the computational effort is large.

There is no systematic way to solve the equations.

The determination of the feedback parameters can be significantly simplified when the invariance of the eigenvalues of a system under a regular transformation is observed and used. The idea is to transform the system into a form which is suitable for the determination of the controller parameters.

Design of a System in Controller Canonical Form

A state-space system in controller canonical form has the following structure:

$$\dot{z}(t)=A_{x}z(t)+b_{c}u(t) \quad z(t_{o}) \text{ initial condition}$$
$$y(t)=c_{c}^{T}z(t)$$

with

$$A_{c}=\begin{bmatrix} 0 & 1 & 0 & . & 0 \\ 0 & 0 & 1 & . & 0 \\ . & . & . & . & . \\ 0 & 0 & 0 & . & 1 \\ -a_{0} & -a_{1} & -a_{2} & . & -a_{n-1} \end{bmatrix}, \quad b_{c}=\begin{bmatrix} 0 \\ \vdots \\ 0 \\ 1 \end{bmatrix}, \quad c_{c}=\begin{bmatrix} c_{1} \\ c_{2} \\ \vdots \\ c_{n} \end{bmatrix}$$

This canonical form has the following properties:

The characteristic polynomial can be directly determined from the last line of A_c, which is

$$P_c(s) = \det(sI - A_c) = a_0 + a_1 s + a_2 s^2 + \cdots + a_{n-1} s^{n-1} + s^n.$$

A system of this structure is always controllable as its controllability matrix according to Equation has always full rank.

The transfer function of the system is immediately given by

$$G(s) = \frac{c_1 + c_2 s + c_3 s^2 + \cdots + c_n s^{n-1}}{a_0 + a_1 s + a_2 s^2 + \cdots + a_{n-1} s^{n-1} + s^n}.$$

The feedback is now defined as

$$u(t) = -f_c^T z(t)$$

with

$$f_c = \begin{bmatrix} f_{c1} \\ f_{c2} \\ \vdots \\ f_{cn} \end{bmatrix}.$$

For the closed-loop system the system matrix is

$$A_c - b_c f_c^T = \begin{bmatrix} 0 & 1 & 0 & . & 0 \\ 0 & 0 & 1 & . & 0 \\ . & . & . & . & . \\ 0 & 0 & 0 & . & 1 \\ -a_0 - f_{c1} & -a_1 - f_{c2} & -a_2 - f_{c3} & . & -a_{n-1} - f_{cn} \end{bmatrix}$$

and its characteristic polynomial is

$$Q(s) = (a_0 + f_{c1}) + (q_1 + f_{c2})s + (a_2 + f_{c3})s^2 + \cdots + (a_{n-1} + f_{cn})s^{n-1} + s^n.$$

Equating this polynomial with the polynomial with the desired poles from Equation one obtains by comparison of the corresponding terms of both polynomials the controller parameters as

$$f_{ci} = p_{i-1} - a_{i-1} \text{ for } i = 1, 2, \ldots, n.$$

In the controller canonical form the calculation of the controller feedback parameters is reduced to the calculation of a simple difference between the coefficients of two polynomials.

Design of a System not in a Canonical Form

In general, when a system is not given in the controller canonical form, one has to transform it by a regular transformation

$$z(t) = Tx\,(t),$$

which brings the system into the desired canonical form according to Equations. The determination of the controller parameters in f_c is performed according to Equation. The feedback law in the original state is, using Equation given by

$$u(t) = -(f_c^T T)\,x(t) = -f^T x(t).$$

Finally, the feedback vector is transformed back to

$$f^T = f_c^T T.$$

The main task in the pole-placement design for systems that are not in controller canonical form, is the determination of the transformation matrix $\boldsymbol{T}$. When the original state equation from Equation is transformed by Equation, one obtains the transformed entities

$$A_c = T\,A\,T^{-1}$$

and

$$b_c = Tb.$$

Equation will be further analysed in the form by right-multiplying with T

$$A_c = T\,A.$$

With the row vectors t_i^T of the matrix T this equation is

$$
\begin{bmatrix}
0 & 1 & 0 & . & 0 \\
0 & 0 & 1 & . & 0 \\
. & . & . & . & . \\
0 & 0 & 0 & . & 1 \\
-a_0 & -a_1 & -a_2 & . & -a_{n-1}
\end{bmatrix}
\begin{bmatrix}
t_1^T \\
t_2^T \\
. \\
t_{n-1}^T \\
t_n^T
\end{bmatrix}
=
\begin{bmatrix}
t_1^T \\
t_2^T \\
. \\
t_{n-1}^T \\
t_n^T
\end{bmatrix}
A
$$

and after multiplication:

$$
\begin{bmatrix}
t_2^T \\
t_3^T \\
. \\
t_n^T \\
-a_0 t_1^T - a_1 t_2^T - \cdots - a_{n-2} t_{n-1}^T - a_{n-1} t_n^T
\end{bmatrix}
=
\begin{bmatrix}
t_1^T A \\
t_2^T A \\
. \\
t_{n-1}^T A \\
t_n^T A
\end{bmatrix}.
$$

From the first $n - 1$ rows of both sides one obtains the recursive relationship

$$t_{i+1}^T = t_i^T A \quad \text{for} \quad i = 1, 2, ..., n - 1,$$

and when t_1^T is known the remaining rows of the matrix T are

$$t_{i+1}^T = t_1^T A^i \quad \text{for} \quad i = 1, 2, ..., n - 1.$$

The first row t_1^T is obtained from Equation, which is also valid. Using the results from Equation, the right-hand side of Equation has the form:

$$b_c = Tb = \begin{bmatrix} t_1^T \\ t_2^T \\ . \\ t_{n-1}^T \\ t_n^T \end{bmatrix} \quad b = \begin{bmatrix} t_1^T b \\ t_1^T b \\ . \\ t_1^T A^{n-2} b \\ t_1^T A^{n-1} b \end{bmatrix}$$

or in transposed form:

$$b_c^T = t_1^T \begin{bmatrix} b & Ab & \cdots & A^{n-2}b & A^{n-1}b \end{bmatrix} = t_1^T S.$$

The matrix S is the controllability matrix from Equation, which has full rank if the system is completely controllable. Under this condition one can obtain the first row by

$$t_1^T = b_c^T S^{-1},$$

which is the last row of the inverse controllability matrix, because b_c is the n-th unit vector.

Chapter 4

PID CONTROL MODES

PID CONTROLLERS EXPLAINED

PID controllers are named after the Proportional, Integral and Derivative control modes they have. They are used in most automatic process control applications in industry. PID controllers can be used to regulate flow, temperature, pressure, level, and many other industrial process variables. This blog reviews the design of PID controllers and explains the P, I and D control modes used in them.

Manual Control

Without automatic controllers, all regulation tasks will have to be done manually. For example: To keep constant the temperature of water discharged from an industrial gas-fired heater, an operator will have to watch a temperature gauge and adjust a fuel gas valve accordingly. If the water temperature becomes too high for some reason, the operator has to close the gas valve a bit – just enough to bring the temperature back to the desired value. If the water becomes too cold, he has to open the gas valve.

Feedback Control

The control task done by the operator is called feedback control, because the operator changes the firing rate based on feedback that he gets from the process via the temperature gauge. Feedback control can be done manually as described here, but it is commonly done automatically.

Control Loop

The operator, valve, process, and temperature gauge forms a control loop. Any change the operator makes to the gas valve affects the temperature which is fed back to the operator, thereby closing the loop.

Automatic Control

To relieve our operator from the tedious task of manual control, we should automate the control loop. This is done as follows:

- Install an electronic temperature measurement device.
- Automate the gas valve by adding an actuator (and perhaps a positioner) to it so that it can be driven electronically.
- Install a controller (in this case a PID controller), and connect it to the electronic temperature measurement and the automated control valve.

A PID controller has a Set Point (SP) that the operator can set to the desired temperature. The Controller's Output (CO) sets the position of the control valve. And the temperature measurement, called the Process Variable (PV) gives the controller its much-needed feedback. The process variable and controller output are commonly transmitted via 4 – 20mA signals, or via digital commands on a Fieldbus.

When everything is up and running, the PID controller compares the process variable to its set point and calculates the difference between the two signals, also called the Error (E).

Then, based on the Error and the PID controller's tuning constants, the controller calculates an appropriate controller output that opens the control valve to the right position for keeping the temperature at the set point. If the temperature should rise above its set point, the controller will reduce the valve position and vice versa.

PID Control

PID controllers have three control modes:

- Proportional Control
- Integral Control
- Derivative Control

Each of the three modes reacts differently to the error. The amount of response produced by each control mode is adjustable by changing the controller's tuning settings.

Proportional Control Mode

The proportional control mode is in most cases the main driving force in a controller. It changes the controller output in proportion to the error. If the error gets bigger, the control action gets bigger. This makes a lot of sense, since more control action is needed to correct large errors.

The adjustable setting for proportional control is called the Controller Gain (K_c). A higher controller gain will increase the amount of proportional control action for a given error. If the controller gain is set too high the control loop will begin oscillating and become unstable. If the controller gain is set too low, it will not respond adequately to disturbances or set point changes.

$$P = K_C \times E$$

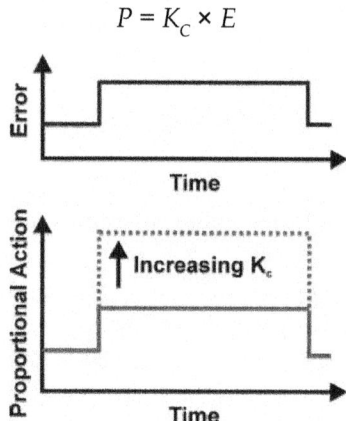

Fig. : Proportional control action.

Adjusting the controller gain setting actually influences the integral and derivative control modes too. That is why this parameter is called *controller gain* and not *proportional gain*.

Proportional Band

While most controllers use controller gain (K_c) as the proportional setting, some controllers use Proportional Band (PB), which is expressed in percent. Table shows the relationship between K_c and PB.

Table : Relationship between K_c and PB.

Controller Gain (K_c)	Proportional Band (PB) %
0.1	1000
0.2	500
0.5	200
1	100
2	50
5	20
10	10

Proportional-only Controller

Proportional controllers are simple to understand and easy to tune. The controller output is simply the output of the proportional control mode, plus a bias. The bias is needed so that the controller can maintain an output (say at 50%) while there is no error (set point = process variable).

$$CO = (K_C \times E) + Bias$$

The use of proportional control alone has a large drawback – offset. Offset is a sustained error that cannot be eliminated by proportional control alone. For example, let's consider controlling the water level in the tank with a proportional-only controller. As long as the flow out of the tank remains constant, the level will remain at its set point.

But, if the operator should increase the flow out of the tank, the tank level will begin to decrease due to the imbalance between inflow and outflow. While the tank level decreases, the error increases and our proportional controller increases the controller output proportional to this error. Consequently, the valve controlling the flow into the tank opens wider and more water flows into the tank.

As the level continues to decrease, the valve continues to open until it gets to a point where the inflow again matches the outflow. At this point the tank level (and error) will remain constant. Because the error remains constant our P-controller will keep its output constant and the control valve will hold its position. The system now remains at balance, but the tank level remains below its set point. This residual sustained error is called Offset.

The effect of a sudden decrease in fuel gas pressure to the process heater described earlier, and the response of a p-only controller. The decrease in fuel-gas pressure reduces the firing rate and the heater outlet temperature decreases. This creates and error to which the controller responds. However, a new balance-point between control action and error is found and the temperature offset is not eliminated by the proportional controller.

Under proportional-only control, the offset will remain until the operator manually changes the bias on the controller's output to remove the offset. This is typically done by putting the controller in manual mode, changing its output manually until the error is zero, and then putting it back in automatic control. It is said that the operator manually "Resets" the controller.

Integral Control Mode

The need for manual reset as described above led to the development of automatic reset or the Integral Control Mode, as we know it today. As long as there is an error present (process variable not at set point), the integral control mode will continuously increment or decrement the controller's output to reduce the error. Given enough time, integral action will drive the controller output far enough to reduce the error to zero.

If the error is large, the integral mode will increment/decrement the controller output fast, if the error is small, the changes will be slower. For a given error, the speed of the integral action is set by the controller's integral time setting (T_I). A large value of T_I (long integral time) results in a slow integral action, and a small value of T_I (short integral time) results in a fast integral action. If the integral time is set too long, the controller will be sluggish, if it is set too short, the control loop will oscillate and become unstable. In the figure, T_S is the control algorithm's execution interval, sometimes called sampling time or scan time.

$$I = I_{previous} + K_C \times E \times \frac{T_S}{T_I}$$

$$CO = K_C \left(\frac{1}{T_I} \int E \, dt \right)$$

Fig. : Integral control action and an integral-only controller's equation.

Most controllers use integral time in minutes as the unit of measure for integral control, but some others use integral time in seconds, integral gain in repeats per minute or repeats per second. Table compares the different integral units of measure.

Table : Units of the integral control mode.

Integral Time		Integral Gain	
Minutes	Seconds	Rep / Min	Rep / Sec
0.05	3	20	0.333
0.1	6	10	0.167
0.2	12	5	0.0833
0.5	30	2	0.0333
1	60	1	0.0167
2	120	0.5	0.00833
5	300	0.2	0.00333
10	600	0.1	0.00167
20	1200	0.05	0.00083

Derivative Control Mode

The third control mode in a PID controller is derivative. Derivative control is rarely used in controlling processes, but it is used often in motion control. For process control, it is not absolutely required, is very sensitive to measurement noise and it makes trial-and-error tuning more difficult.

The derivative control mode produces an output based on the rate of change of the error. Derivative mode is sometimes called Rate. The derivative mode produces more control action if the error changes at a faster rate. If there is no change in the error, the derivative action is zero. The derivative mode has an adjustable setting called Derivative Time (T_D). The larger the derivative time setting, the more derivative action is produced. A derivative time setting of zero effectively turns off this mode. If the derivative time is set too long, oscillations will occur and the control loop will run unstable. Again T_S is the controller's execution interval.

Proportional + Integral + Derivative Controller

Commonly called the PID controller, its controller output is made up of the sum of the proportional, integral, and derivative control actions. There are other configurations too.

PID control provides more control action sooner than what is possible with P or PI control. This reduces the effect of a disturbance, and shortens the time it takes for the level to return to its set point.

ANATOMY OF A FEEDBACK CONTROL SYSTEM

Here is the classic block diagram of a process under PID Control.

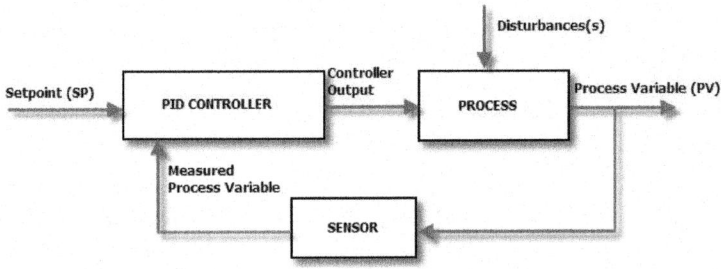

What's going on this diagram? The **Setpoint** (SP) is the value that we **want** the process to be.

For example, the temperature control system in our house may have a SP of 22°C. This means that

"we want the heating and cooling process in our house to achieve a steady temperature of as close to 22°C as possible"

The PID controller looks at the setpoint and compares it with the actual value of the **Process Variable (PV)**. Back in our house, the box of electronics that is the PID controller in our Heating and Cooling system looks at the value of the temperature sensor in the room and sees how close it is to 22°C.

If the SP and the PV are the same – then the controller is a very happy little box. It doesn't have to do anything, it will set its output to zero.

However, if there is a disparity between the SP and the PV we have an error and corrective action is needed. In our house this will either be cooling or heating depending on whether the PV is higher or lower than the SP respectively.

Let's imagine the temperature PV in our house is higher than the SP. It is too hot. The air-con is switched on and the temperature drops.

The sensor picks up the lower temperature, feeds that back to the controller, the controller sees that the "temperature error" is not as great because the PV (temperature) has dropped and the air con is turned down a little.

This process is repeated until the house has cooled down to 22°C and there is no error.

Then a disturbance hits the system and the controller has to kick in again.

In our house the disturbance may be the sun beating down on the roof, raising the temperature of the air inside.

So that's a really, really basic overview of a simple feedback control system. Sounds dead simple eh?

Understanding the controller

Unfortunately, in the real world we need a controller that is a bit more complicated than the one described above, if we want top performance form our loops. To understand why, we will be doing some "thought experiments" where we are the controller.

When we have gone through these thought experiments we will appreciate why a PID algorithm is needed and why/how it works to control the process.

We will be using the analogy of changing lanes on a freeway on a windy day. We are the driver, and therefore the controller of the process of changing the car's position.

Here's the Block Diagram we used before, with the labels changed to represent the car-on-windy-freeway control loop.

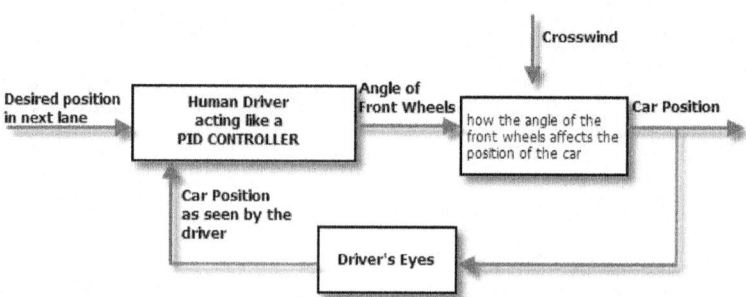

Notice how important closing the feedback loop is. If we removed the feedback loop we would be in "open loop control", and would have to control the car's position with our eyes closed!

Thankfully we are under "Closed loop control" -using our eyes for position feedback.

As we saw in the house-temperature example the controller takes the both the PV and SP signals, which it then puts through a black box to calculate a controller output. That controller output is sent to an actuator which moves to actually control the process.

We are interested here in what the black box actually does, which is that it applies 1, 2 or 3 calculations to the SP and Measured PV signals. These calculations, called the "Modes of Control" include:

- Proportional (P)
- Integral (I)
- Derivative (D)

Under The Hood Of The PID Controller

Here's a simplified block diagram of what the PID controller does:

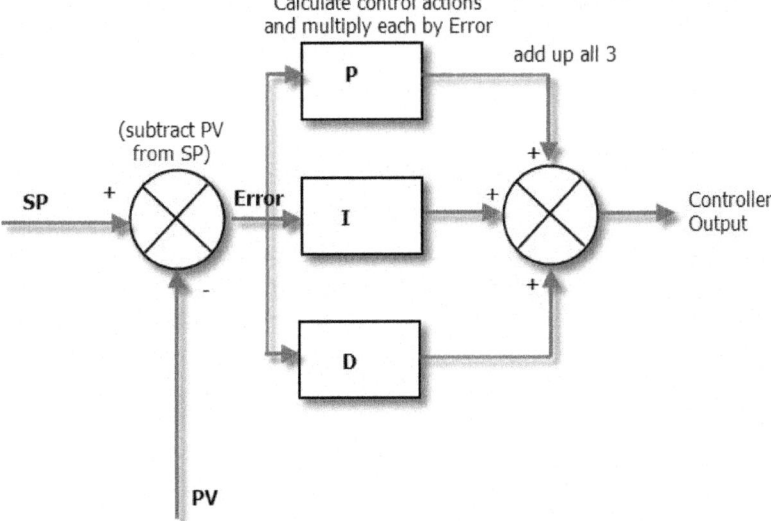

It is really very simple in operation. The PV is subtracted from the SP to create the Error. The error is simply multiplied by one, two or all of the calculated P, I and D actions (depending which ones are turned on). Then the resulting "error x control actions" are added together and sent to the controller output.

These 3 modes are used in different combinations:

P – Sometimes used

PI - Most often used

PID – Sometimes used

PD – rare as hen's teeth but can be useful for controlling servomotors.

Derivatives

Go into the control room of a process plant and ask the operator:

"What's the derivative of reactor 4's pressure?"

And the response will typically be:

"Bugger off smart arse!"

However go in and ask:

"What's the rate of change of reactor 4's pressure?"

And the operator will examine the pressure trend and say something like:

"About 5 PSI every 10 minutes"

He's just performed calculus on the pressure trend! (don't tell him though or he'll want a pay raise)

So derivative is just a mathematical term meaning rate-of-change. That's all there is to it.

Integrals without the Math

Is it any wonder that so many people run scared from the concept of integrals and integration, when this is a typical definition?

Here's a plain English definition:

The integral of a signal is the sum of all the instantaneous values that the signal has been, from whenever you started counting until you stop counting.

So if you are to plot your signal on a trend and your signal is sampled every second, and let's say you are measuring temperature. If you were to superimpose the integral of the signal over the first 5 seconds – it would look like this:

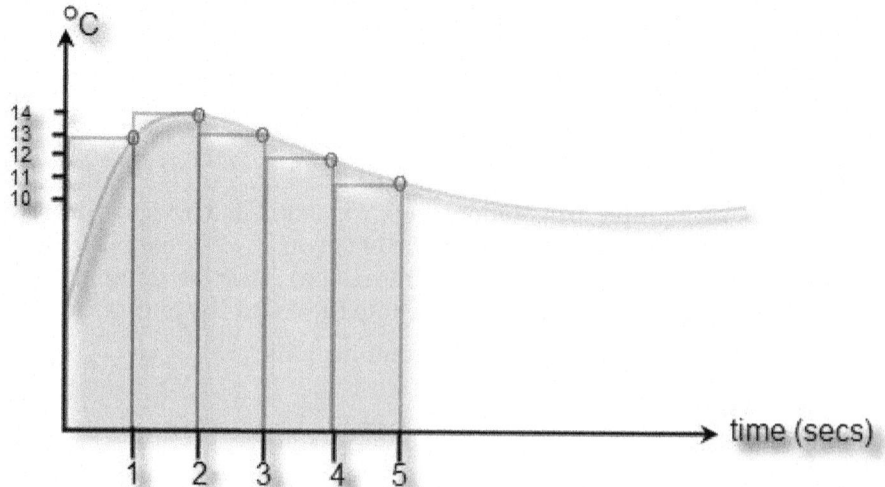

The green line is your temperature, the red circles are where your control system has sampled the temperature and the blue area is the integral of the temperature signal. It is the sum of the 5 temperature values over the time period that you are interested in. In numerical terms it is the sum of the areas of each of the blue rectangles:

$$(13 \times 1) + (14 \times 1) + (13 \times 1) + (12 \times 1) + (11 \times 1) = 63 \,°C \, s$$

The curious units (degrees Celsius x seconds) are because we have to multiply a temperature by a time – but the units aren't important.

As you can probably remember from school –the integral turns out to be the area under the curve. When we have real world systems, we actually get an approximation to the area under the curve, which as you can see from the diagram gets better, the faster we sample.

Proportional Control

Here's a diagram of the controller when we have enabled only P control:

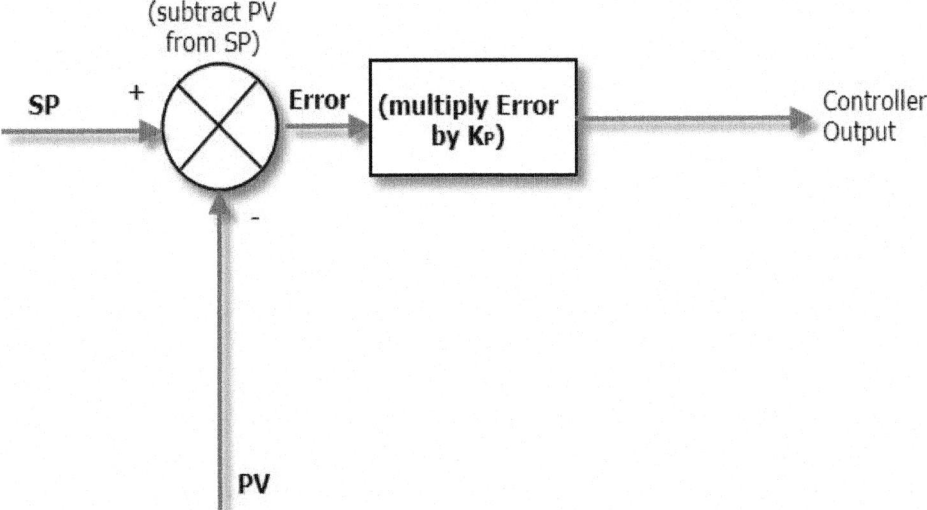

In Proportional Only mode, the controller simply multiplies the Error by the Proportional Gain (K_p) to get the controller output.

The Proportional Gain is the setting that we tune to get our desired performance from a "P only" controller.

A match made in heaven: The P + I Controller

If we put Proportional and Integral Action together, we get the humble PI controller. The Diagram below shows how the algorithm in a PI controller is calculated.

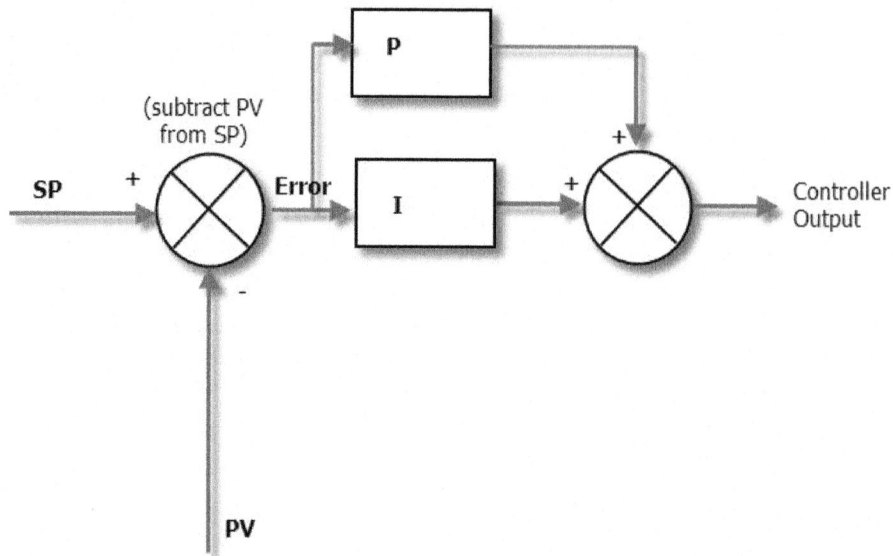

The tricky thing about Integral Action is that it will really screw up your process unless you know **exactly** how much Integral action to apply.

A good PID Tuning technique will calculate exactly how much Integral to apply for your specific process - but how is the Integral Action adjusted in the first place?

Adjusting the Integral Action

The way to adjust how much Integral Action you have is by adjusting a term called "minutes per repeat". Not a very intuitive name is it?

So where does this strange name come from? It is a measure of how long it will take for the Integral Action to match the Proportional Action.

In other words, if the output of the proportional box on the diagram above is 20%, the repeat time is the time it will take for the output of the Integral box to get to 20% too.

And the important point to note is that the "bigger" integral action, the quicker it will get this 20% value. That is, it will take fewer minutes to get there, so the "minutes per repeat" value will be smaller.

In other words the smaller the "minutes per repeat" is the bigger the integral action.

To make things a bit more intuitive, a lot of controllers use an alternative unit of "repeats per minute" which is obviously the inverse of "minutes per repeat".

The nice thing about "repeats per minute" is that the bigger it is - the bigger the resulting Integral action is.

Derivative Action – predicting the future

OK, so the combination of P and I action seems to cover all the bases and do a pretty good job of controlling our system. That is the reason that PI controllers are the most prevalent. They do the job well enough and keep things simple. Great.

But engineers being engineers are always looking to tweak performance.

They do this in a PID loop by adding the final ingredient: Derivative Action.

So adding derivative action can allow you to have bigger P and I gains and still keep the loop stable, giving you a faster response and better loop performance.

If you think about it, Derivative action improves the controller action because it predicts what is yet to happen by projecting the current rate of change into the future. This means that it is not using the current measured value, but a future measured value.

The units used for derivative action describe how far into the future you want to look. *i.e.* If derivative action is 20 seconds, the derivative term will project the current rate of change 20 seconds into the future.

The big problem with D control is that if you have noise on your signal (which looks like a bunch of spikes with steep sides) this confuses the hell out of the algorithm. It looks at the slope of the noise-spike and thinks:

"Holy crap! This process is changing quickly, lets pile on the D Action!!!"

And your control output jumps all over the place, messing up your control.

Of course you can try and filter the noise out, but my advice is that, unless PI control is really slow, don't worry about switching D on.

PID CONTROL AND DERIVATIVE ON MEASUREMENT

Like the PI controller, the Proportional-Integral-Derivative (PID) controller computes a controller output (CO) signal for the final control element every sample time T.

The PID controller is a "three mode" controller. That is, its activity and performance is based on the values chosen for three tuning parameters, one each nominally associated with the proportional, integral and derivative terms.

The **PI controller** is a reasonably straightforward equation with two adjustable tuning parameters. The number of different ways that commercial vendors can implement the PI form is fairly limited, and they all provide the same performance if properly tuned.

With the addition of a third adjustable tuning parameter, the number of algorithm permutations increases markedly. And there are even **different forms of the PID equation**itself. This creates added challenges for controller design and tuning.

Here we focus on what a derivative is, how it is computed, and what it means for control. We also explore why derivative on measurement is widely recommended for industrial practice.

The Dependent, Ideal PID Form

A popular way vendors express the dependent, ideal PID controller is:

$$CO = CO_{bias} + Kc \cdot e(t) + \frac{Kc}{Ti} \int e(t)dt + Kc \cdot Td \frac{de(t)}{dt}$$

Where:

CO   = controller output signal (**the wire out**)

CO_{bias} = controller bias; set by **bumpless transfer**

e(t)   = current controller error, defined as SP – PV

SP   = set point

PV   = measured process variable (**the wire in**)

Kc   = controller gain, a tuning parameter

Ti   = reset time, a tuning parameter

Td   = derivative time, a tuning parameter

The derivative mode of the PID controller is an additional and separate term added to the end of the equation that considers the derivative (or rate of change) of the error as it varies over time.

The Contribution of the Derivative Term

The **proportional term** considers *how far* PV is from SP at any instant in time. Its contribution to the CO is based on the size of e(t) only at time t. As e(t) grows or shrinks, the influence of the proportional term grows or shrinks immediately and proportionately.

The **integral term** addresses *how long* and how far PV has been away from SP. The integral term is continually summing e(t). Thus, even a small error, if it persists, will have a sum total that grows over time and the influence of the integral term will similarly grow.

A derivative describes how steep a curve is. More properly, a derivative describes the slope or the rate of change of a signal trace at a particular point in time. Accordingly, the derivative term in the PID equation above considers *how fast*, or the rate at which, error is changing at the current moment.

Derivative on PV is Opposite but Equal

While the proportional and integral terms of the PID equation are driven by the controller error, e(t), the derivative computation in many commercial implementations should be based on the value of PV itself.

The derivative of e(t) is mathematically identical to the negative of the derivative of PV everywhere except when set point changes. And when set point changes, derivative on error results in an undesirable control action called *derivative kick*.

Math Note: the mathematical defense that "derivative of e(t) equals the negative derivative of PV when SP is constant" considers that, since

$$e(t) = SP - PV,$$

the equation below follows. That is, derivative of error equals derivative of set point minus process variable. The derivative of a constant is zero, so when SP is constant, mathematically, the derivative (or slope or rate of change) of the controller error equals the derivative (or slope or rate of change) of the measured process variable, PV, except the sign is opposite.

$$\frac{de(t)}{dt} = \frac{d(\overset{0}{\cancel{SP}} - PV)}{dt} = -\frac{dPV}{dt}$$

The figures below provide a visual appreciation that the derivative of e(t) is the negative of the derivative of PV.

The top plot shows the measured PV trace after a set point step. The bottom plot shows the e(t) = SP − PV trace for the same event.

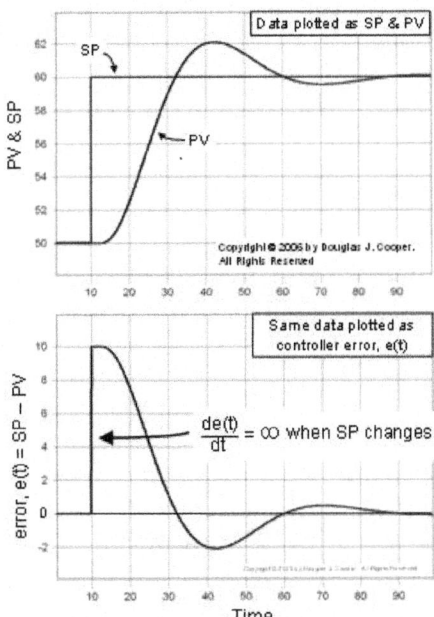

If we compare the two plots after the SP step at time t = 10, we see that the PV trace in the upper plot is an exact reflection of the e(t) trace in the lower plot. The PV trace ascends, peaks and then settles, while in a reflected pattern, the e(t) trace descends, dips and then settles.

Mathematically, this "mirror image" of trace shapes means that the derivatives (or slopes or rates of change) are the same everywhere after the SP step, except they are opposite in sign.

Derivative on PV Used in Practice

While the shape of e(t) and PV are opposite but equal everywhere after the set point step, there is an important difference at the moment the SP changes. The lower plot shows a vertical spike in e(t) at this moment. There is no corresponding spike in the PV plot.

The derivative (or slope) of a vertical spike in the theoretical world approaches infinity. In the real world it is at least a very big number. If Td is large enough to provide any meaningful weight to the derivative term, this huge derivative value will cause a large and sudden manipulation in CO. This large manipulation in CO, referred to as *derivative kick*, is almost always undesirable.

As long as loop sample time, T, **is properly specified**, the PV trace will follow a gradual and continuous response, avoiding the dramatic vertical spike evident in the e(t) trace.

Because derivative on e(t) is identical to derivative on PV at all times except when the SP changes, and when the set point does change, derivative on error provides information we don't want our controller to use, we substitute the "math note" equation in the yellow box above to obtain the PID with derivative on measurement controller:

$$CO = CO_{bias} + Kc \cdot e(t) + \frac{Kc}{Ti} \int e(t)dt - Kc \cdot Td\frac{dPV}{dt}$$

Derivative on PV Does Not "Kick"

The first set point steps to the left in the plot below show loop performance when PID with derivative on error is used. The set points steps to the right present the identical scenario except that PID with derivative on measurement is used.

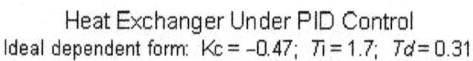

Heat Exchanger Under PID Control
Ideal dependent form: Kc = –0.47; Ti = 1.7; Td = 0.31

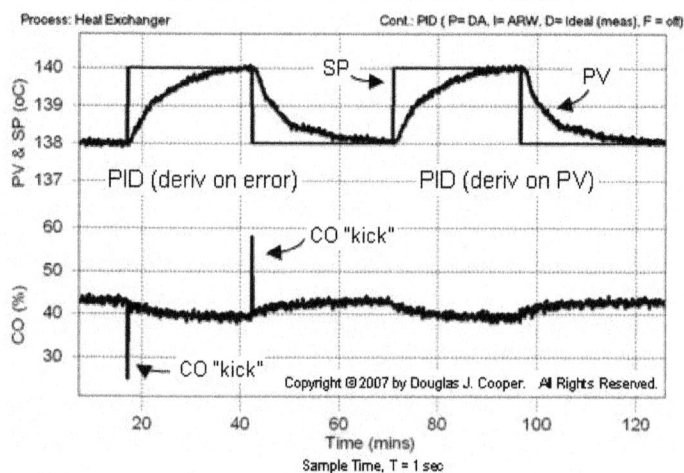

The "kick" that dominates the CO trace when derivative on error is used is rather dramatic and somewhat unsettling. While it exists for only a brief moment and does not impact performance in this example, we should not assume this will always be true. In any event, such action will eventually take a toll on mechanical final control elements.

We recommend that derivative on measured PV be used if our vendor provides the option (fortunately most do). The tuning values remain the same for both algorithms.

Understanding Derivative Action

A rapidly changing PV has a steep slope and this yields a large derivative. This is true regardless of whether a dynamic event has just begun or if it has been underway for some time.

The derivative dPV/dt describes the slope or "steepness" of PV during a process response.

Early in the response, the slope is large and positive when the PV trace is increasing rapidly. When PV is decreasing, the derivative (slope) is negative. And when the PV goes through a peak or a trough, there is a moment in time when the derivative is zero.

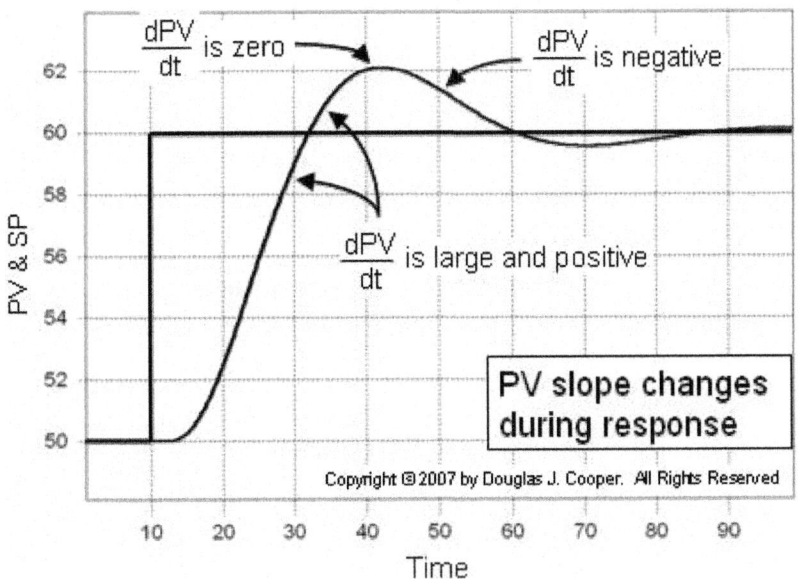

To understand the impact of this changing derivative, let's assume for discussion that:

Controller gain, Kc, is positive.

Derivative time, Td (always positive) is large enough to provide meaningful weight to the derivative term. After all, if Td is very small, the derivative term has little influence, regardless of the slope of the PV.

The negative sign in front of the derivative term of the PID with derivative on measurement controller (and given the above assumptions) means that the impact on CO from the derivative term will be opposite to the sign of the slope:

$$CO = CO_{bias} + Kc \cdot e(t) + \frac{Kc}{Ti}\int e(t)dt - Kc \cdot Td\frac{dPV}{dt}$$

Thus, when dPV/dt is large and positive, the derivative term has a large influence and seeks to decrease CO.

Conversely, when dPV/dt is negative, the derivative term seeks to increase CO.

It is interesting to note that the derivative term does not consider whether PV is heading toward or away from the set point (whether e(t) is positive or negative). The only consideration is whether PV is heading up or down and how quickly.

The result is that derivative action seeks to inhibit rapid movements in the PV. This could be an especially useful characteristic when seeking to dampen the oscillations in PV that integral action tends to magnify. Unfortunately, the potential benefit comes with a price.

BUMPLESS CONTROL TRANSFER BETWEEN MANUAL AND PID CONTROL

Model Description

This example shows how to achieve bumpless control transfer when switching from manual control to PID control. We use the PID Controller block in Simulink® to control a first-order process with dead-time.

We start by opening the model.

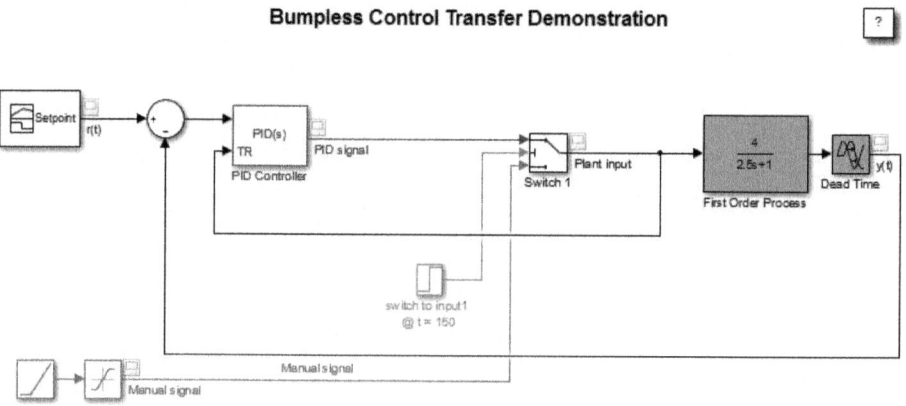

Fig. : Simulink model of PID control with bumpless transfer.

To open this model, type sldemo_bumpless in a MATLAB® terminal.

The PID Controller has been tuned with saturation ignored using the PID tuner of Simulink® Control Design™.

The controlled plant is a first-order process with dead-time described by

$$P(s) = \frac{4}{2.5s + 1} e^{-s}$$

For several operational reasons, the engineers decided to start the control process in an open-loop manner by feeding the plant input with a saturating ramp signal to drive the output of the plant slowly to a desired steady-state value of 40. A control transfer is scheduled to happen a $t = 150$. This transition between open-loop control and closed-loop control therefore involves two control phases of operation:

1. Manual: A saturated ramp signal feeds the plant input during start-up until $t = 150$.

2. Automatic: A PID controller will engage the plant at $t = 150$, and must take over the process without introducing bumps at the plant input.

To support smooth control transition, the PID Controller block supports two modes of operation: a tracking mode and a control mode. In control mode, the PID Controller block operates as an ordinary PID controller. In tracking mode, however, the block has an extra input that allows the PID block to adjust its internal state by changing its integrator output so that the block output tracks a prescribed signal feeding this extra input port.

To achieve bumpless control transfer, the PID Controller block must be in tracking mode when the plant is in the manual control phase (open-loop control), and in control mode when the plant is in the automatic control phase (closed-loop control).

Configuring the Block for Tracking Mode

To activate signal tracking, go to the PID Advanced tab in the block's dialog; select Enable tracking mode, and specify the gain Kt. The inverse of this gain is the time constant of the tracking loop.

Fig. : Enabling the tracking mode of the PID Controller block.

Once tracking mode is enabled, the block has a second input port denoted by TR. Internally this new port is wired as shown under mask:

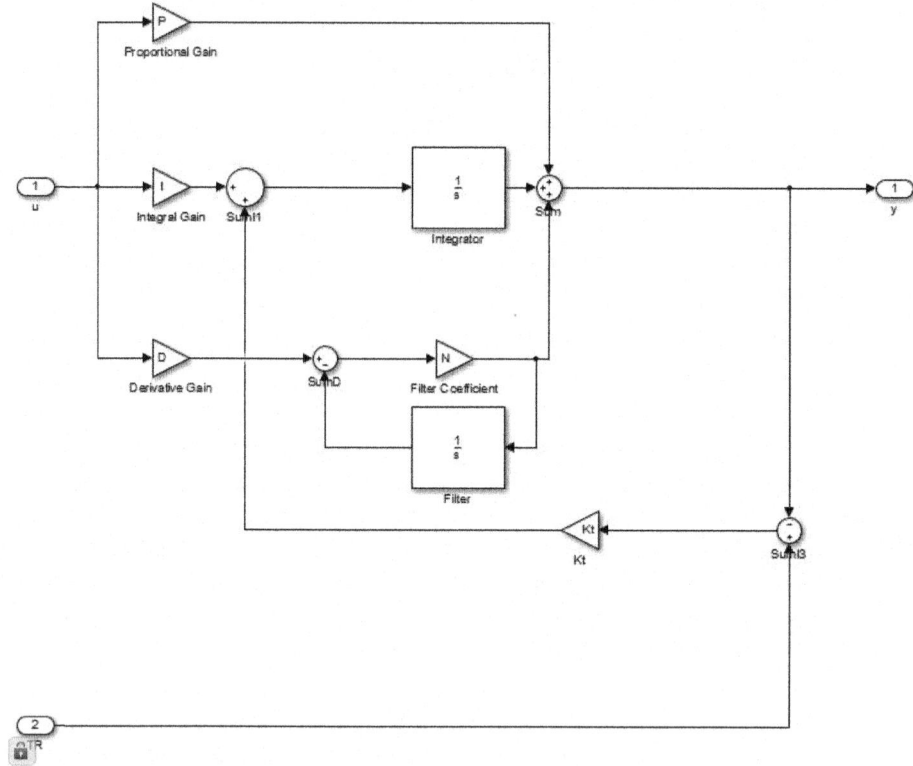

Fig. : Under-mask view of the PID Controller block with tracking mode.

Setting Up the Switching Mechanism

In addition to enabling tracking mode for the PID Controller block, a switching mechanism is needed to achieve the control transfer. Switch 1 determines which signal feeds the plant input and feeds the tracking port of the PID Controller block.

At time t = 0, Switch 1 directs the manual control signal to the plant input and the tracking port of the PID Controller block. This allows the output of the PID Controller block to track the manual control signal during the manual phase by adjusting the PID Controller's internal integrator. When control transfer occurs, therefore, the PID Controller output will be approximately the same as the manual control signal.

At time t = 150, Switch 1 switches, directing the output of the PID Controller block to the plant input and the PID Controller block's tracking input. The PID Controller block now tracks its own output, which is equivalent to control mode.

Simulating the Bumpless Control Transfer

The setpoint signal and the closed-loop response of the model are shown in Figure below.

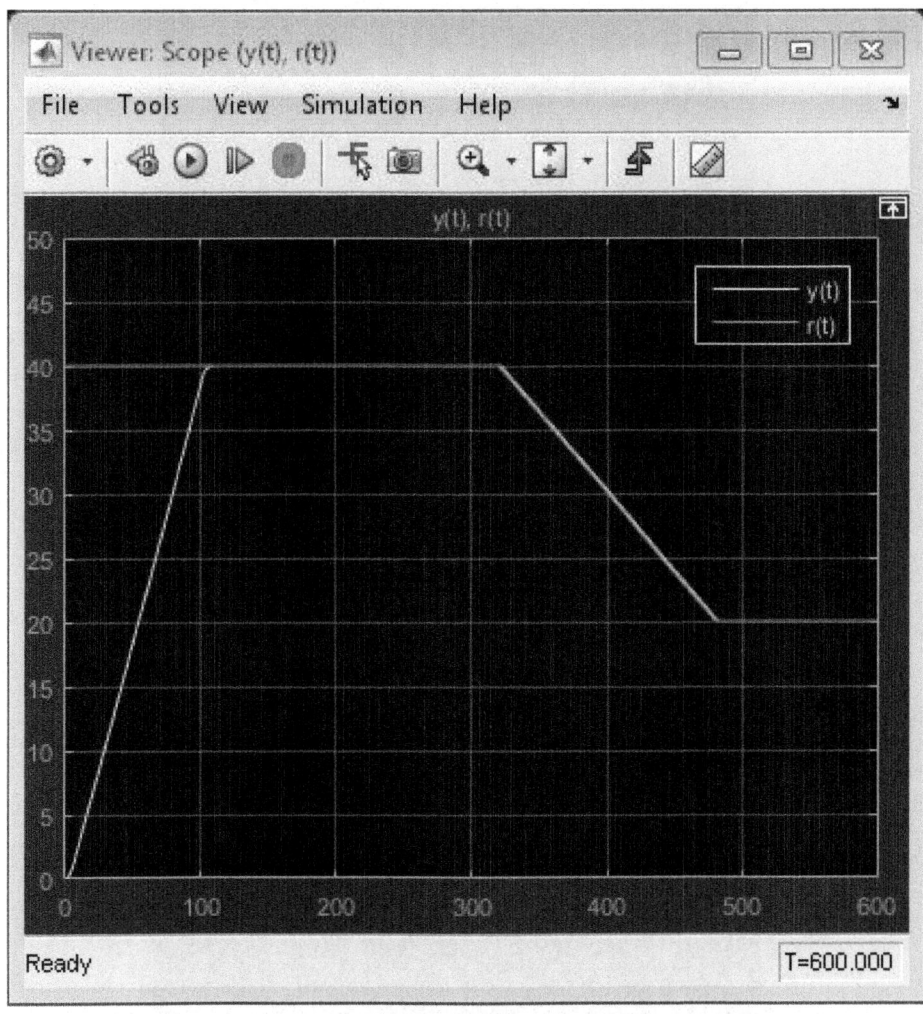

Fig. : Setpoint vs. measured output.

The measured output tracks the Setpoint profile without any output bumps at the time of switching (t = 150). To investigate this further, the plant input, control signals.

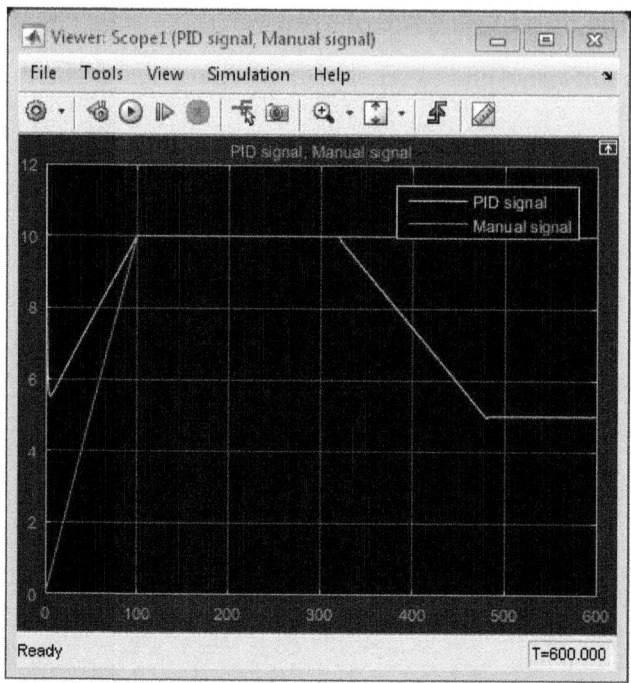

Fig. : Control signal switching.

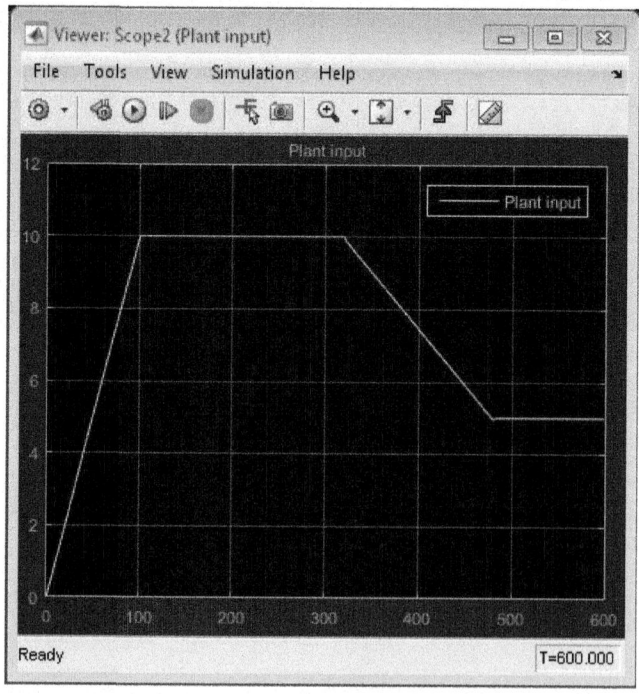

Fig. : Plant input.

The switching instance, the plant input has experienced no step changes (bumps), and therefore the control transfer happens in smooth bumpless fashion as intended.

To see the significance of the bumpless transfer setup, consider the case where tracking mode is not used. In this case, the following setup is obtained:

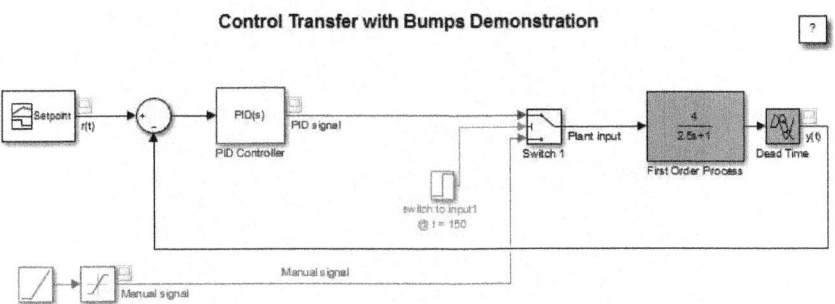

Fig. : Simulink model of PID control with no bumpless transfer.

To open this model, type sldemo_bumplessno in a MATLAB terminal.

The performance in the absence of an appropriate bumpless control transfer strategy.

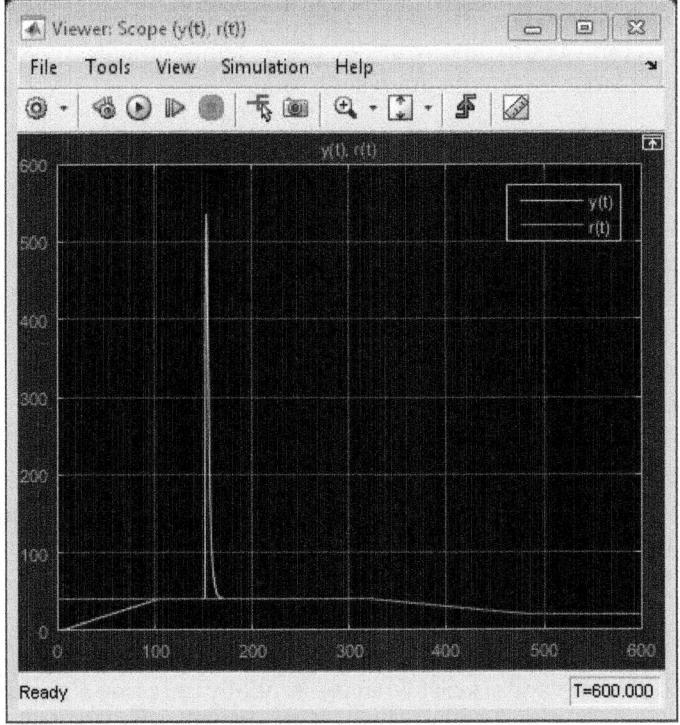

Fig. : Setpoint vs. measured output.

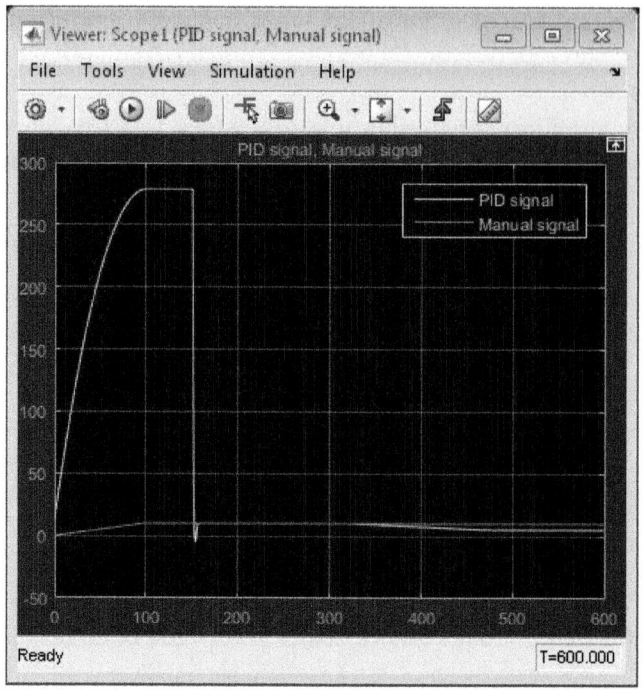

Fig. : Control signal switching.

The PID controller to float while the plant is under manual control can result in undesirable large transients upon switching.

INTEGRAL ACTION AND PI CONTROL

Like the **P-Only controller**, the Proportional-Integral (PI) algorithm computes and transmits a controller output (CO) signal every sample time, T, to the final control element (*e.g.*, valve, variable speed pump). The computed CO from the PI algorithm is influenced by the controller tuning parameters and the controller error, e(t).

PI controllers have two tuning parameters to adjust. While this makes them more challenging to tune than a P-Only controller, they are not as complex as the three parameter **PID controller**.

Integral action enables PI controllers to eliminate offset, a major weakness of a P-only controller. Thus, PI controllers provide a balance of complexity and capability that makes them by far the most widely used algorithm in process control applications.

The PI Algorithm

While different vendors cast what is essentially the **same algorithm in different forms**, here we explore what is variously described as the dependent, ideal, continuous, position form:

$$CO = CO_{bias} + Kc \cdot e(t) + \frac{Kc}{Ti} \int e(t) dt$$

Where:

CO = controller output signal (the wire out)

CO_{bias} = controller bias or null value; set by bumpless transfer as explained below

e(t) = current controller error, defined as SP – PV

SP = set point

PV = measured process variable (the wire in)

Kc = controller gain, a tuning parameter

Ti = reset time, a tuning parameter

The integral mode of the controller is the last term of the equation. Its function is to integrate or continually sum the controller error, e(t), over time.

Some things we should know about the reset time tuning parameter, Ti:

It provides a separate weight to the integral term so the influence of integral action can be independently adjusted.

It is in the denominator so smaller values provide a larger weight to (i.e. increase the influence of) the integral term.

It has units of time so it is always positive.

Function of the Proportional Term

As with the P-Only controller, the proportional term of the PI controller, Kc·e(t), adds or subtracts from CO_{bias} based on the size of controller error e(t) at each time t.

As e(t) grows or shrinks, the amount added to CO_{bias} grows or shrinks immediately and proportionately. The past history and current trajectory of the controller error have no influence on the proportional term computation.

The plot below illustrates this idea for a set point response. The error used in the proportional calculation is shown on the plot:

At time t = 25 min, e(25) = 60 – 56 = 4

At time t = 40 min, e(40) = 60 – 62 = –2

Recalling that controller error e(t) = SP – PV, rather than viewing PV and SP as separate traces as we do above, we can compute and plot e(t) at each point in time t.

Below is the identical data to that above only it is recast as a plot of e(t) itself. Notice that in the plot above, PV = SP = 50 for the first 10 min, while in the error plot below, e(t) = 0 for the same time period.

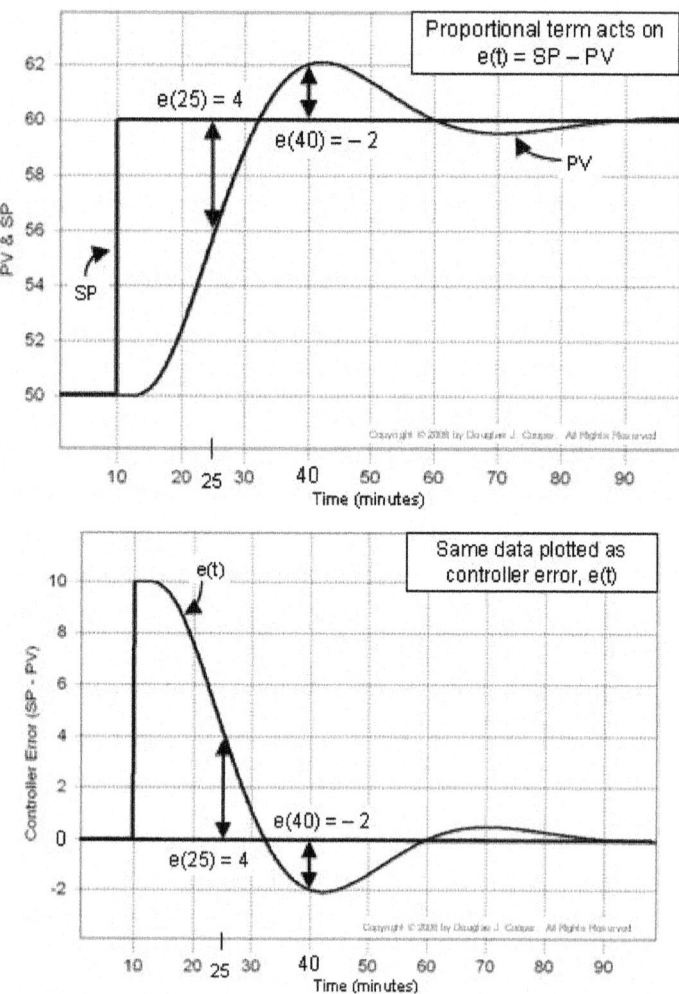

This plot is useful as it helps us visualize how controller error continually changes size and sign as time passes.

Function of the Integral Term

While the proportional term considers the current size of e(t) only at the time of the controller calculation, the integral term considers the history of the error, or how long and how far the measured process variable has been from the set point over time.

Integration is a continual summing. Integration of error over time means that we sum up the complete controller error history up to the present time, starting from when the controller was first switched to automatic.

Controller error is e(t) = SP – PV. In the plot below, the integral sum of error is computed as the shaded areas between the SP and PV traces.

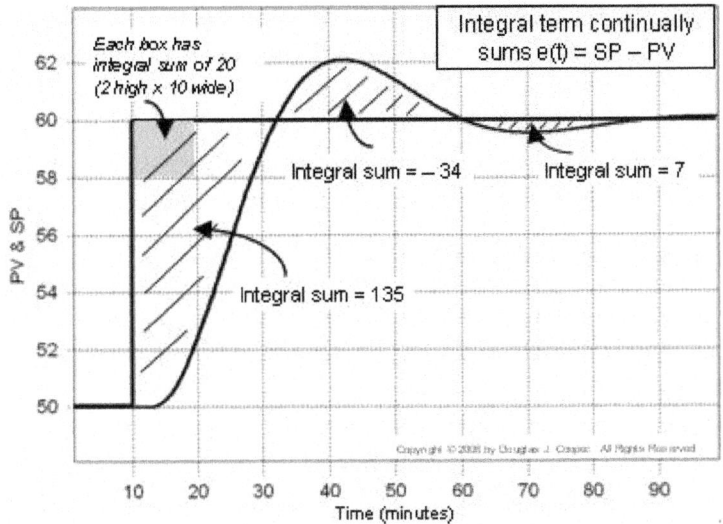

Each box in the plot has an integral sum of 20 (2 high by 10 wide). If we count the number of boxes (including fractions of boxes) contained in the shaded areas, we can compute the integral sum of error.

So when the PV first crosses the set point at around t = 32, the integral sum has grown to about 135. We write the integral term of the PI controller as:

$$\frac{Kc}{Ti}\int_0^{32} e(t)dt = \frac{Kc}{Ti}(135)$$

Since it is controller error that drives the calculation, we get a direct view the situation from a controller error plot as shown below:

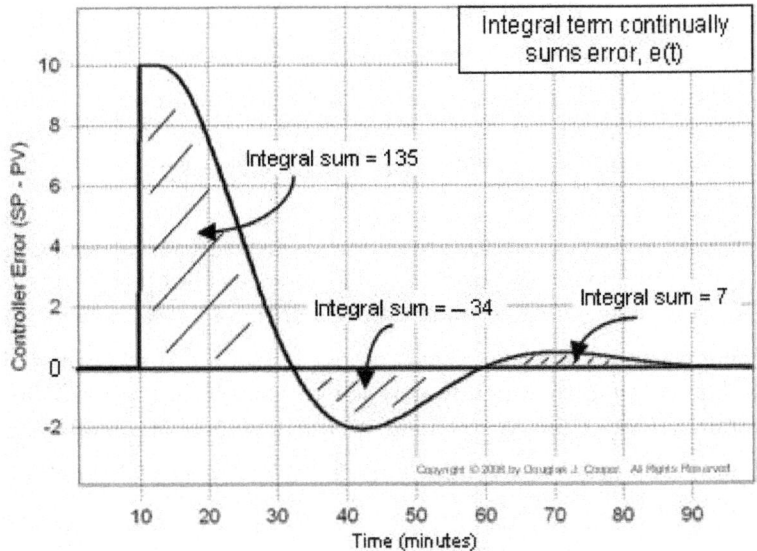

Note that the integral of each shaded portion has the same sign as the error. Since the integral sum starts accumulating when the controller is first put in automatic, the total integral sum grows as long as e(t) is positive and shrinks when it is negative.

At time t = 60 min on the plots, the integral sum is 135 – 34 = 101. The response is largely settled out at t = 90 min, and the integral sum is then 135 – 34 + 7 = 108.

Integral Action Eliminates Offset

The previous sentence makes a subtle yet very important observation. The response is largely complete at time t = 90 min, yet the integral sum of all error is not zero.

In this example, the integral sum has a final or residual value of 108. It is this residual value that enables integral action of the PI controller to eliminate offset.

Most processes under P-only control experience offset during normal operation. Offset is a sustained value for controller error (*i.e.*, PV does not equal SP at steady state).

We recognize from the P-Only controller:

$$CO = CO_{bias} + Kc \cdot e(t)$$

that CO will always equal CO_{bias} unless we add or subtract something from it.

The only way we have something to add or subtract from CO_{bias} in the P-Only equation above is if e(t) is not zero. It e(t) is not steady at zero, then PV does not equal SP and we have offset.

However, with the PI controller:

$$CO = CO_{bias} + Kc \cdot e(t) + \frac{Kc}{Ti} \int_0^{32} e(t)dt$$

we now know that the integral sum of error can have a final or residual value after a response is complete. This is important because it means that e(t) can be zero, yet we can still have something to add or subtract from CO_{bias} to form the final controller output, CO.

So as long as there is any error (as long as e(t) is not zero), the integral term will grow or shrink in size to impact CO. The changes in CO will only cease when PV equals SP (when e(t) = 0) for a sustained period of time.

At that point, the integral term can have a residual value as just discussed. This residual value from integration, when added to CO_{bias}, essentially creates a new overall bias value that corresponds to the new level of operation.

In effect, integral action continually resets the bias value to eliminate offset as operating level changes.

Challenges of PI Control

There are challenges in employing the PI algorithm:

The two tuning parameters interact with each other and their influence must be balanced by the designer.

The integral term tends to increase the oscillatory or rolling behaviour of the process response.

Because the two tuning parameters interact with each other, it can be challenging to arrive at "best" tuning values. The value and importance of our **design and tuning recipe** increases as the controller becomes more complex.

Initializing the Controller for Bumpless Transfer

When we switch any controller from manual mode to automatic (from open loop to closed loop), we want the result to be uneventful. That is, we do not want the switchover to cause abrupt control actions that impact or disrupt our process

We achieve this desired outcome at switchover by initializing the controller integral sum of error to zero. Also, the set point and controller bias value are initialized by setting:

1. SP equal to the current PV

2. CO_{bias} equal to the current CO

With the integral sum of error set to zero, there is nothing to add or subtract from CO_{bias} that would cause a sudden change in the current controller output. With the set point equal to the measured process variable, there is no error to drive a change in our CO. And with the controller bias set to our current CO value, we are prepared by default to maintain current operation.

Thus, when we switch from manual mode to automatic, we have "bumpless transfer" with no surprises. This is a result everyone appreciates.

Reset Time Versus Reset Rate

Different vendors cast their **control algorithms in slightly different forms**. Some use proportional band rather than controller gain. Also, some use reset rate, Tr, instead of reset time. These are simply the inverse of each other:

$$Tr = 1/Ti$$

No matter how the tuning parameters are expressed, the PI algorithms are all equally capable.

PID CONTROLLER DESIGN

We will introduce a simple yet versatile feedback compensator structure, the Proportional-Integral-Derivative (PID) controller. The effect of each of the pid parameters on the closed-loop dynamics and demonstrate how to use a PID controller to improve the system performance.

PID Overview

We will consider the following unity feedback system:

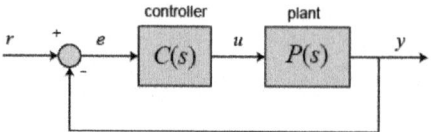

The output of a PID controller, equal to the control input to the plant, in the time-domain is as follows:

$$u(t) = K_p e(t) + K_i \int e(t)dt + K_p \frac{de}{dt}$$

First, let's take a look at how the PID controller works in a closed-loop system using the schematic shown above. The variable (e) represents the tracking error, the difference between the desired input value (r) and the actual output (y). This error signal (e) will be sent to the PID controller, and the controller computes both the derivative and the integral of this error signal. The control signal (u) to the plant is equal to the proportional gain (K_p) times the magnitude of the error plus the integral gain (K_i) times the integral of the error plus the derivative gain (K_d) times the derivative of the error.

This control signal (u) is sent to the plant, and the new output (y) is obtained. The new output (y) is then fed back and compared to the reference to find the new error signal (e). The controller takes this new error signal and computes its derivative and its integral again, ad infinitum.

The transfer function of a PID controller is found by taking the Laplace transform of Equation.

$$K_p + \frac{K_i}{s} + K_d s = \frac{K_d s^2 + K_p s + K_i}{s}$$

K_p = Proportional gain K_i = Integral gain K_d = Derivative gain

We can define a PID controller in MATLAB using the transfer function directly, for example:

```
Kp = 1;
Ki = 1;
Kd = 1;
s = tf('s');
C = Kp + Ki/s + Kd*s
C =
s^2 + s + 1
-----------
     s
Continuous-time transfer function.
```

Alternatively, we may use MATLAB's pid controller object to generate an equivalent continuous-time controller as follows:

```
C = pid(Kp,Ki,Kd)
C =

            1
Kp + Ki * --- + Kd * s

            s
with Kp = 1, Ki = 1, Kd = 1
Continuous-time PID controller in parallel form.
```

Let's convert the pid object to a transfer function to see that it yields the same result as above:

```
tf(C)
ans =
    s^2 + s + 1
    -----------

         s
Continuous-time transfer function.
```

The Characteristics of P, I, and D Controllers

A proportional controller (K_p) will have the effect of reducing the rise time and will reduce but never eliminate the steady-state error. An integral control (K_i) will have the effect of eliminating the steady-state error for a constant or step input, but it may make the transient response slower. A derivative control (K_d) will have the effect of increasing the stability of the system, reducing the overshoot, and improving the transient response.

The effects of each of controller parameters, K_p, K_d, and K_i on a closed-loop system are summarized in the table below.

Cl Response	Rise Time	Overshoot	Settling Time	S-S Error
Kp	Decrease	Increase	Small Change	Decrease
Ki	Decrease	Increase	Increase	Eliminate
Kd	Small Change	Decrease	Decrease	No Change

Note that these correlations may not be exactly accurate, because K_p, K_i, and K_d are dependent on each other. In fact, changing one of these variables can change the effect of the other two. For this reason, the table should only be used as a reference when you are determining the values for K_i, K_p and K_d.

Example Problem

Suppose we have a simple mass, spring, and damper problem.

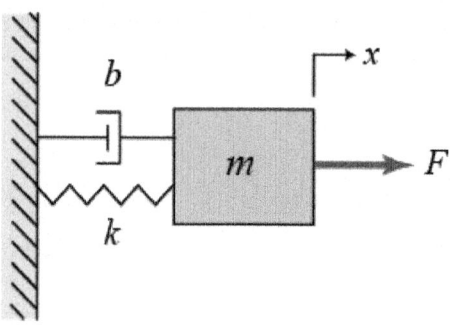

The modeling equation of this system is

$$M\ddot{x} + b\dot{x} + kx = F$$

Taking the Laplace transform of the modeling equation, we get

$$Ms^2 X(s) + bsX(s) + kX(s) = F(s)$$

The transfer function between the displacement $X(s)$ and the input $F(s)$ then becomes

$$\frac{X(s)}{F(s)} = \frac{1}{Ms^2 + bs + k}$$

Let

```
M = 1 kg
b = 10 N s/m
k = 20 N/m
F = 1 N
```

Plug these values into the above transfer function

$$\frac{X(s)}{F(s)} = \frac{1}{s^2 + 10s + 20}$$

The goal of this problem is to show you how each of K_p, K_i and K_d contributes to obtain

```
Fast rise time
Minimum overshoot
No steady-state error
```

Open-Loop Step Response

Let's first view the open-loop step response. Create a new m-file and run the following code:

```
s = tf('s');
P = 1/(s^2 + 10*s + 20);
step(P)
```

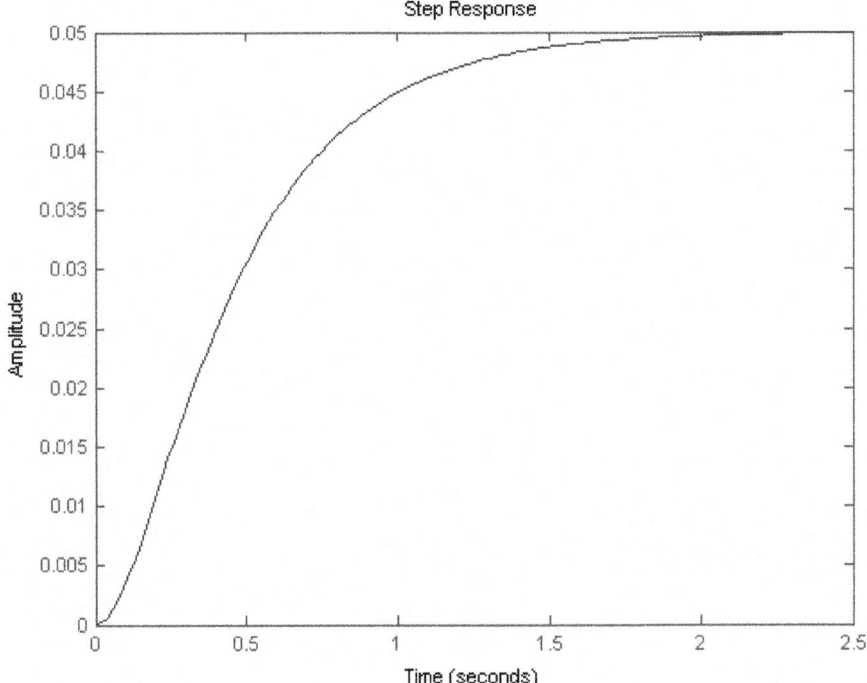

The DC gain of the plant transfer function is 1/20, so 0.05 is the final value of the output to an unit step input. This corresponds to the steady-state error of 0.95, quite large indeed. Furthermore, the rise time is about one second, and the settling time is about 1.5 seconds. Let's design a controller that will reduce the rise time, reduce the settling time, and eliminate the steady-state error.

Proportional Control

From the table shown above, we see that the proportional controller (Kp) reduces the rise time, increases the overshoot, and reduces the steady-state error.

The closed-loop transfer function of the above system with a proportional controller is:

$$\frac{X(s)}{F(s)} = \frac{K_p}{s^2 + 10s + (20 + K_p)}$$

Let the proportional gain (K_p) equal 300 and change the m-file to the following:

```
Kp = 300;
C = pid(Kp)
T = feedback(C*P,1)
t = 0:0.01:2;
step(T,t)
```

```
C =
Kp = 300
P-only controller.
T =      300
       ----------------
       s^2 + 10 s + 320
Continuous-time transfer function.
```

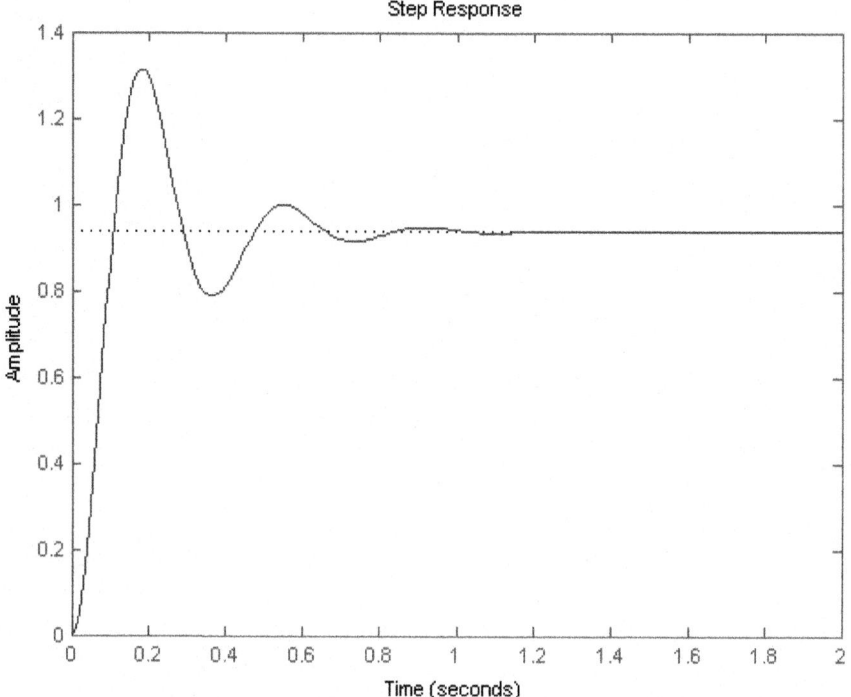

The above plot shows that the proportional controller reduced both the rise time and the steady-state error, increased the overshoot, and decreased the settling time by small amount.

Proportional-Derivative Control

Now, let's take a look at a PD control. From the table shown above, we see that the derivative controller (Kd) reduces both the overshoot and the settling time. The closed-loop transfer function of the given system with a PD controller is:

$$\frac{X(s)}{F(s)} = \frac{K_d s + K_p}{s^2 + (10 + K_d)s + (20 + K_p)}$$

Let K_p equal 300 as before and let K_d equal 10. Enter the following commands into an m-file and run it in the MATLAB command window.

```
Kp = 300;
Kd = 10;
C = pid(Kp,0,Kd)
T = feedback(C*P,1)
t = 0:0.01:2;
step(T,t)
C =
Kp + Kd * s
with Kp = 300, Kd = 10
Continuous-time PD controller in parallel form.
T =        10 s + 300
       ----------------
       s^2 + 20 s + 320
Continuous-time transfer function.
```

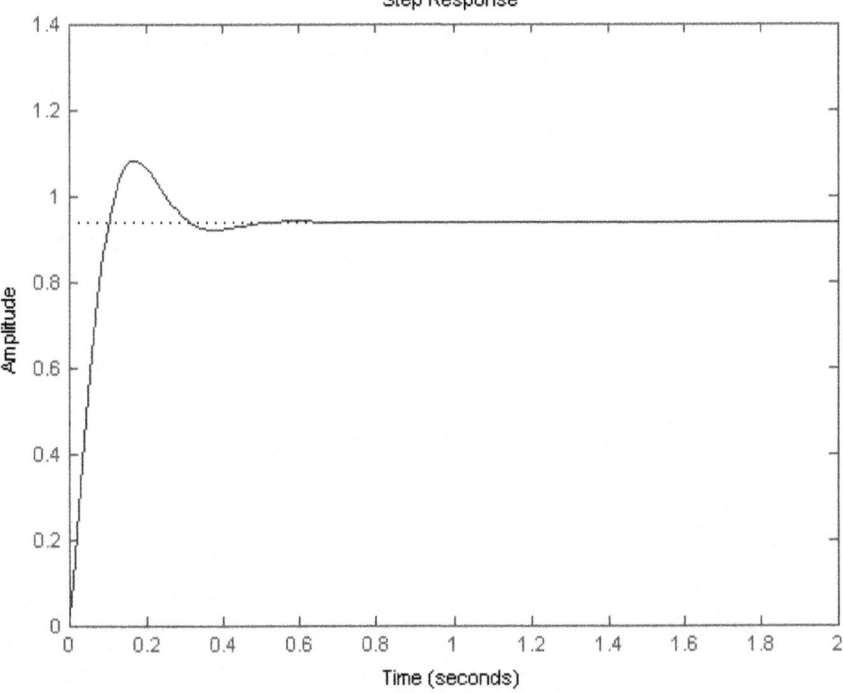

This plot shows that the derivative controller reduced both the overshoot and the settling time, and had a small effect on the rise time and the steady-state error.

Proportional-Integral Control

Before going into a PID control, let's take a look at a PI control. From the table, we see that an integral controller (Ki) decreases the rise time, increases both the overshoot and the settling time, and eliminates the steady-state error. For the given system, the closed-loop transfer function with a PI control is:

$$\frac{X(s)}{F(s)} = \frac{K_d s + K_p}{s^3 + 10e^2 + (20 + K_p s + K_i)}$$

Let's reduce the K_p to 30, and let K_i equal 70. Create an new m-file and enter the following commands.

```
Kp = 30;
Ki = 70;
C = pid(Kp,Ki)
T = feedback(C*P,1)
t = 0:0.01:2;
step(T,t)
C =
            1
Kp + Ki *  ---
            s
with Kp = 30, Ki = 70
Continuous-time PI controller in parallel form.
T =    30 s + 70
      -------------------------
      s^3 + 10 s^2 + 50 s + 70
Continuous-time transfer function.
```

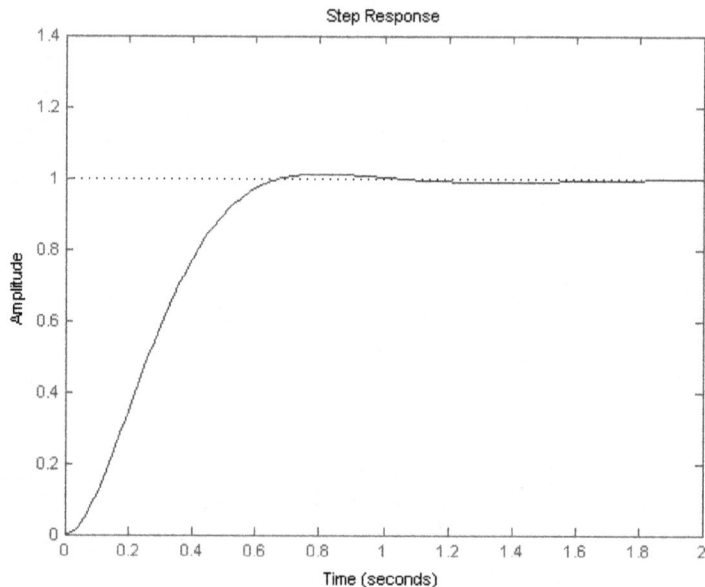

Run this m-file in the MATLAB command window, and you should get the following plot. We have reduced the proportional gain (Kp) because the integral controller also reduces the rise time and increases the overshoot as the proportional controller does (double effect). The above response shows that the integral controller eliminated the steady-state error.

Proportional-Integral-Derivative Control

Now, let's take a look at a PID controller. The closed-loop transfer function of the given system with a PID controller is:

$$\frac{X(s)}{F(s)} = \frac{K_d s^2 + K_p s + K_i}{s^3 + (10 + K_d)s^2 + (20 + K_p)s + K_i}$$

After several trial and error runs, the gains $K_p = 350$, $K_i = 300$, and $K_d = 50$ provided the desired response. To confirm, enter the following commands to an m-file and run it in the command window. You should get the following step response.

```
Kp = 350;
Ki = 300;
Kd = 50;
C = pid(Kp,Ki,Kd)
T = feedback(C*P,1);
t = 0:0.01:2;
step(T,t)
C =             1
Kp + Ki * --- + Kd * s

          s
with Kp = 350, Ki = 300, Kd = 50
Continuous-time PID controller in parallel form.
```

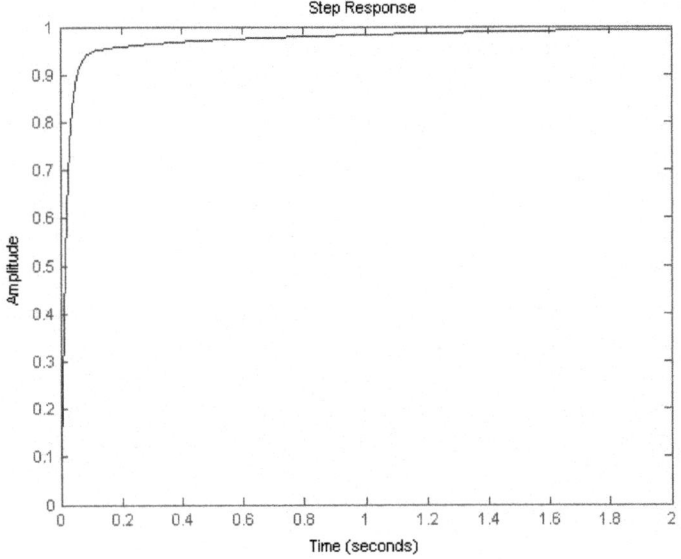

Now, we have obtained a closed-loop system with no overshoot, fast rise time, and no steady-state error.

General Tips for Designing a PID Controller

When you are designing a PID controller for a given system, follow the steps shown below to obtain a desired response.

1. Obtain an open-loop response and determine what needs to be improved
2. Add a proportional control to improve the rise time
3. Add a derivative control to improve the overshoot
4. Add an integral control to eliminate the steady-state error
5. Adjust each of Kp, Ki, and Kd until you obtain a desired overall response.

Lastly, please keep in mind that you do not need to implement all three controllers (proportional, derivative, and integral) into a single system, if not necessary. For example, if a PI controller gives a good enough response (like the above example), then you don't need to implement a derivative controller on the system. Keep the controller as simple as possible.

Automatic PID Tuning

MATLAB provides tools for automatically choosing optimal PID gains which makes the trial and error process described above unnecessary. You can access the tuning algorithm directly using pidtune or through a nice graphical user interface (GUI) usingpidtool.

The MATLAB automated tuning algorithm chooses PID gains to balance performance (response time, bandwidth) and robustness (stability margins). By default the algorthm designs for a 60 degree phase margin.

Let's explore these automated tools by first generating a proportional controller for the mass-spring-damper system by entering the following commands:

```
pidtool(P,'p')
```

The pidtool GUI window, like that shown below, should appear.

Notice that the step response shown is slower than the proportional controller we designed by hand. Now click on the Show Parameters button on the top right. As expected the proportional gain constant, Kp, is lower than the one we used, Kp = 94.85 < 300.

We can now interactively tune the controller parameters and immediately see the resulting response int he GUI window. Try dragging the resposne time slider to the right to 0.14s, as shown in the figure below. The response does indeeed speed up, and we can see Kp is now closer to the manual value. We can also see all the other performance and robustness parameters for the system. Note that the phase margin is 60 degrees, the default for pidtool and generally a good balance of robustness and performance.

PID Tuner
Design mode: Basic
Form Parallel
Type P
Plot: Step
Response: Reference tracking
Show baseline
Show parameters
Amplitude
Time (seconds)
Tuned response
Interactive tuning
Response time: 0.244 seconds
Slower
Faster
Close
Help

PID Tuner
Design mode: Basic
Form Parallel
Type P
Plot: Step
Response: Reference tracking
Hide parameters
Amplitude
Time (seconds)
Tuned response
Controller parameters
Tuned
Kp 232.8481
Ki
Kd
Tf
Performance and robustness
Tuned
Rise time (seconds) 0.0846
Settling time (seconds) 0.704
Overshoot (%) 35.3
Peak 1.25
Gain margin (rad/s) Inf @ Inf
Phase margin (rad/s) 37.8 @ 14.3
Closed-loop stability Stable
Interactive tuning
Response time: 0.14 seconds
Slower
Faster
Close
Help

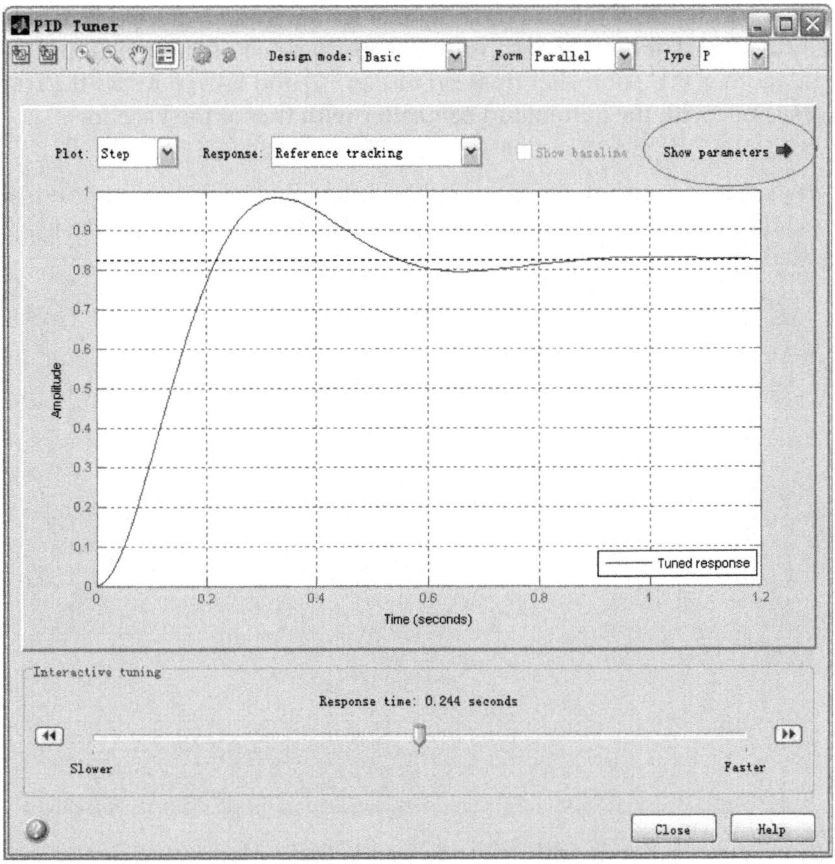

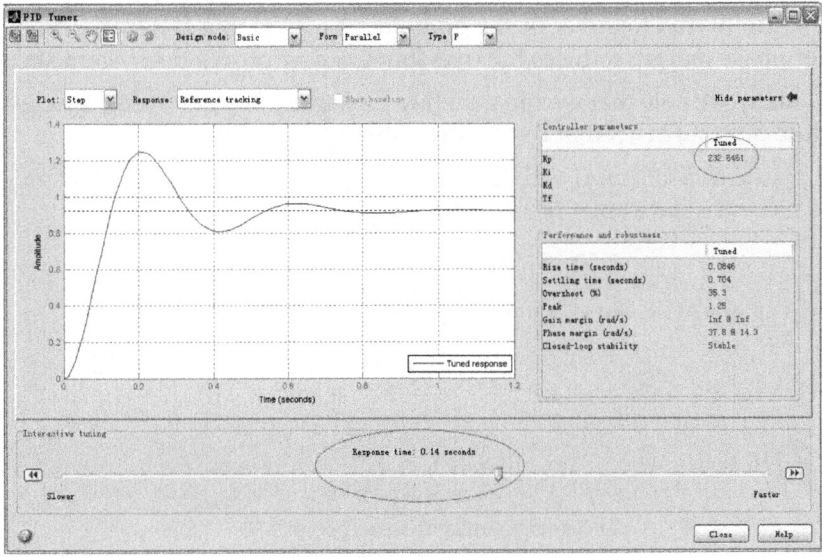

Now let's try designing a PID controller for our system. By specifying the previously designed or (baseline) controller, C, as the second parameter, pidtool will design another PID controller (instead of P or PI) and will compare the response of the system with the automated controller with that of the baseline.

```
pidtool(P,C)
```

We see in the output window that the automated controller responds slower and exhibits more overshoot than the baseline. Now choose the Design Mode: Extended option at the top, which reveals more tuning parameters.

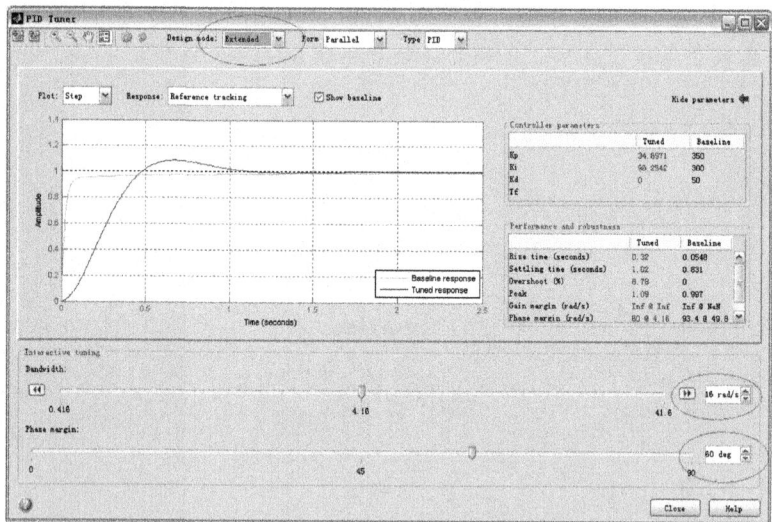

Now type in Bandwidth: 32 rad/s and Phase Margin: 90 deg to generate a controller similar in performance to the baseline. Keep in mind that a higher bandwidth (0 dB crossover of the open-loop) results in a faster rise time, and a higher phase margin reduces the overshoot and improves the system stability.

Finally we note that we can generate the same controller using the command line tool pidtune instead of the pidtool GUI

```
opts = pidtuneOptions('CrossoverFrequency',32,'PhaseMargin',90);[C,
info] = pidtune(P, 'pid', opts)
  C =

              1
  Kp + Ki * --- + Kd * s
              s

  with Kp = 320, Ki = 169, Kd = 31.5
  Continuous-time PID controller in parallel form.
  info =

                    Stable: 1
        CrossoverFrequency: 32
               PhaseMargin: 90
```

ROOT LOCUS CONTROLLER DESIGN

We will introduce the root locus, show how to create it using MATlAB, and demonstrate how to design feedback controllers using the root locus that satisfy certain performance criteria.

Closed-Loop Poles

The root locus of an (open-loop) transfer function $H(s)$ is a plot of the locations (locus) of all possible closed-loop poles with proportional gain K and unity feedback.

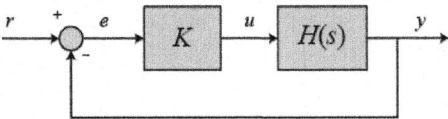

The closed-loop transfer function is:

$$\frac{Y(s)}{R(s)} = \frac{KH(s)}{1+KH(s)}$$

and thus the poles of the closed-loop poles of the closed-loop system are values of s such that $1 + KH(s) = 0$.

If we write $H(s) = b(s)/\,a(s)$, then this equation has the form:

$$a(s) + Kb(s) = 0$$

$$\frac{a(s)}{K} + b(s) = 0$$

Let n = order of $a(s)$ and m = order of $b(s)$ (the order of a polynomial is the highest power of s that appears in it).

We will consider all positive values of K. In the limit as $K \to 0$, the poles of the closed-loop system are $a(s) = 0$ or the poles of $H(s)$. In the limit as $K \to \infty$, the poles of the closed-loop system are $b(s) = 0$ or the zeros of $H(s)$.

No matter what we pick K to be, the closed-loop system must always have n poles, where n is the number of poles of $H(s)$. The root locus must have n branches, each branch starts at a pole of $H(s)$ and goes to a zero of $H(s)$. If $H(s)$ has more poles than zeros (as is often the case), $m < n$ and we say that $H(s)$ has zeros at infinity. In this case, the limit of $H(s)$ as $s \to \infty$ is zero. The number of zeros at infinity is $n - m$, the number of poles minus the number of zeros, and is the number of branches of the root locus that go to infinity (asymptotes).

Since the root locus is actually the locations of all possible closed-loop poles, from the root locus we can select a gain such that our closed-loop system will perform the way we want. If any of the selected poles are on the right half plane, the closed-loop system will be unstable. The poles that are closest to the imaginary axis have the greatest influence on the closed-loop response, so even though the system has three or four poles, it may still act like a second or even first order system depending on the location(s) of the dominant pole(s).

Plotting the Root Locus of a Transfer Function

Consider an open-loop system which has a transfer function of

$$H(s) = \frac{Y(s)}{U(s)} = \frac{s+7}{s(s+5)(s+15)(s+20)}$$

How do we design a feedback controller for the system by using the root locus method? Say our design criteria are 5% overshoot and 1 second rise time. Make a MATLAB file called rl.m. Enter the transfer function, and the command to plot the root locus:

```
s = tf('s');
sys = (s + 7)/(s*(s + 5)*(s + 15)*(s + 20));
rlocus(sys)
axis([-22 3 -15 15])
```

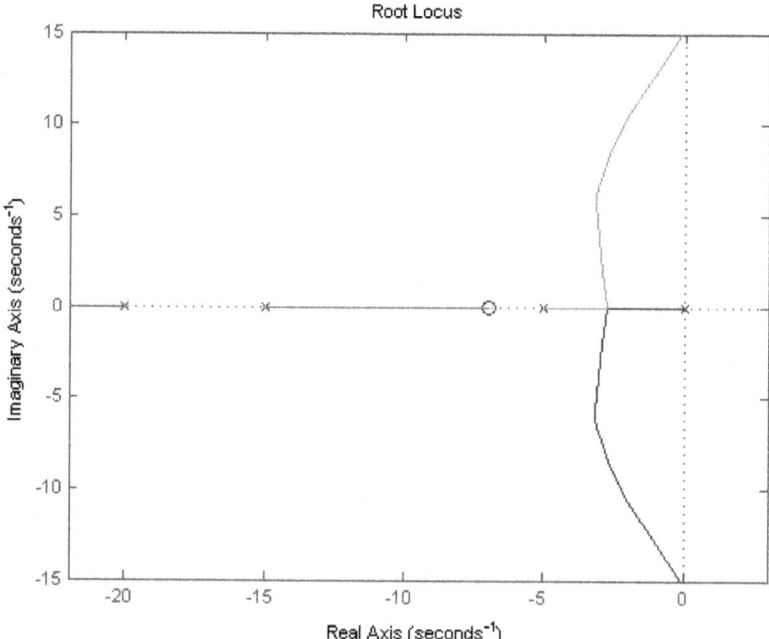

Choosing a Value of K from the Root Locus

The plot above shows all possible closed-loop pole locations for a pure proportional controller. Obviously not all of those closed-loop poles will satisfy our design criteria, To determine what part of the locus is acceptable, we can use the command sgrid(Zeta,Wn) to plot lines of constant damping ratio and natural frequency. Its two arguments are the damping ratio (ζ) and natural frequency (ω_n) [these may be vectors if you want to look at a range of acceptable values]. In our problem, we need an overshoot less than 5% (which means a damping ratio ζ of greater than 0.7) and a rise time of 1 second (which means a natural frequency ω_n greater than 1.8). Enter the following in the MATLAB command window:

```
Zeta = 0.7;
Wn = 1.8;
sgrid(Zeta,Wn)
```

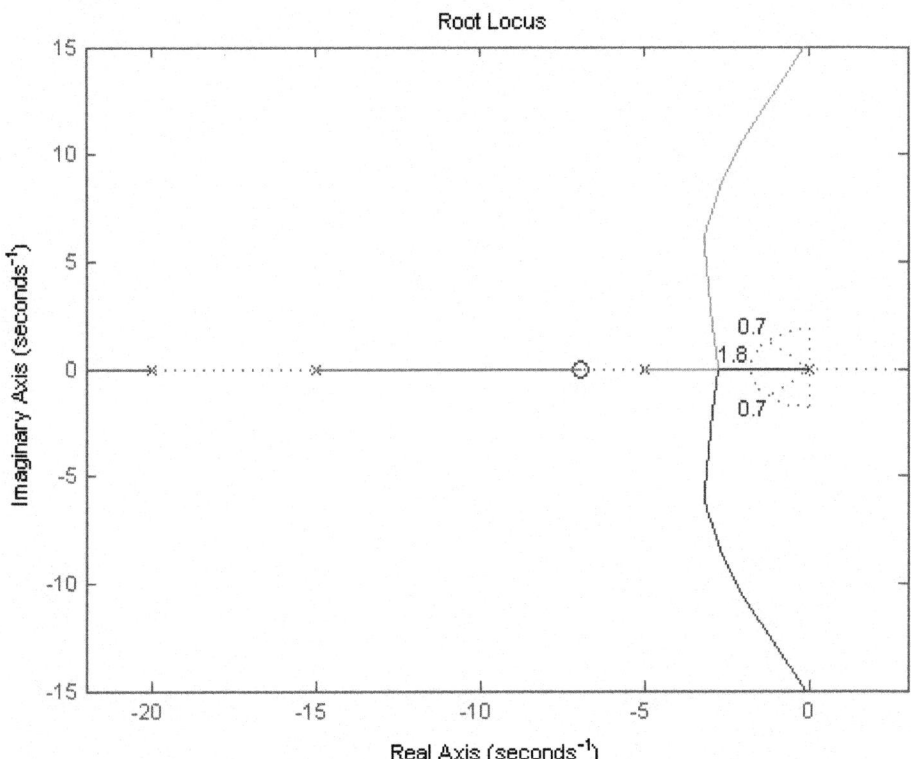

On the plot above, the two dotted lines at about a 45 degree angle indicate pole locations with $\zeta = 0.7$; in between these lines, the poles will have $\zeta > 0.7$ and outside of the lines $\zeta < 0.7$. The semicircle indicates pole locations with a natural frequency $\omega_n = 1.8$; inside the circle, $\omega_n < 1.8$ and outside the circle $\omega_n > 1.8$.

Going back to our problem, to make the overshoot less than 5%, the poles have to be in between the two white dotted lines, and to make the rise time shorter than 1 second, the poles have to be outside of the white dotted semicircle. So now we know only the part of the locus outside of the semicircle and in between the two lines are acceptable. All the poles in this location are in the left-half plane, so the closed-loop system will be stable.

From the plot above we see that there is part of the root locus inside the desired region. So in this case, we need only a proportional controller to move the poles to the desired region. You can use the `rlocfind` command in MATLAB to choose the desired poles on the locus:

$$[k,poles] = rlocfind(sys)$$

Click on the plot the point where you want the closed-loop pole to be. You may want to select the points indicated in the plot below to satisfy the design criteria.

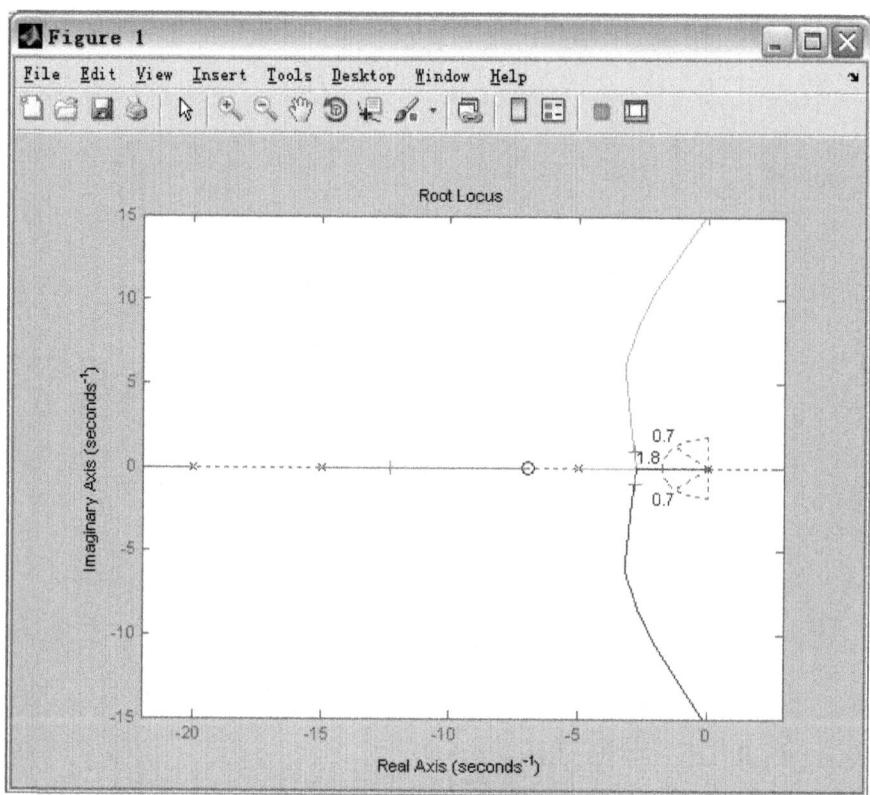

Note that since the root locus may have more than one branch, when you select a pole, you may want to find out where the other pole (poles) are. Remember they will affect the response too. From the plot above, we see that all the poles selected (all the "+" signs) are at reasonable positions. We can go ahead and use the chosen K as our proportional controller.

Closed-Loop Response

In order to find the step response, you need to know the closed-loop transfer function. You could compute this using the rules of block diagrams, or let MATLAB do it for you (there is no need to enter a value for K if the `rlocfind` command was used):

```
K = 350;
sys_cl = feedback(K*sys,1)
sys_cl =
              350 s + 2450
    ---------------------------------------
    s^4 + 40 s^3 + 475 s^2 + 1850 s + 2450
Continuous-time transfer function.
```

The two arguments to the function `feedback` are the numerator and denominator of the open-loop system. You need to include the proportional gain that you have chosen. Unity feedback is assumed.

If you have a non-unity feedback situation, look at the help file for the MATLAB function `feedback`, which can find the closed-loop transfer function with a gain in the feedback loop.

Check out the step response of your closed-loop system:

```
step(sys_cl)
```

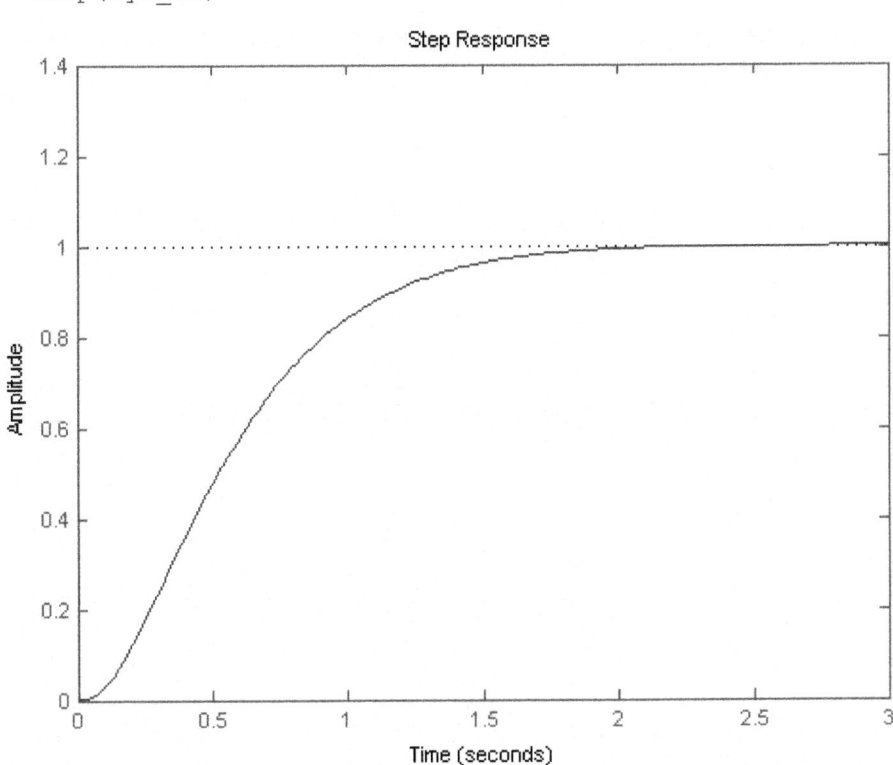

As we expected, this response has an overshoot less than 5% and a rise time less than 1 second.

Using SISOTOOL for Root Locus Design

Another way to complete what was done above is to use the interactive MATLAB GUI called `sisotool`. Using the same model as above, first define the plant, $H(s)$.

```
s = tf('s');
plant = (s + 7)/(s*(s + 5)*(s + 15)*(s + 20));
```

The `sisotool` function can be used for analysis and design. In this case, we will focus on using the Root Locus as the design method to improve the step response of the plant. To begin, type the following into the MATLAB command window:

```
sisotool(plant)
```

The following window should appear. To start, select the tab labeled Graphical Tuning. Within this window, turn off Plot 2 and make sure Plot 1 is the Root Locus and verify that Open Loop 1 is selected. Finally, click the button labeled Show Design Plot to bring up the tunable Root Locus plot.

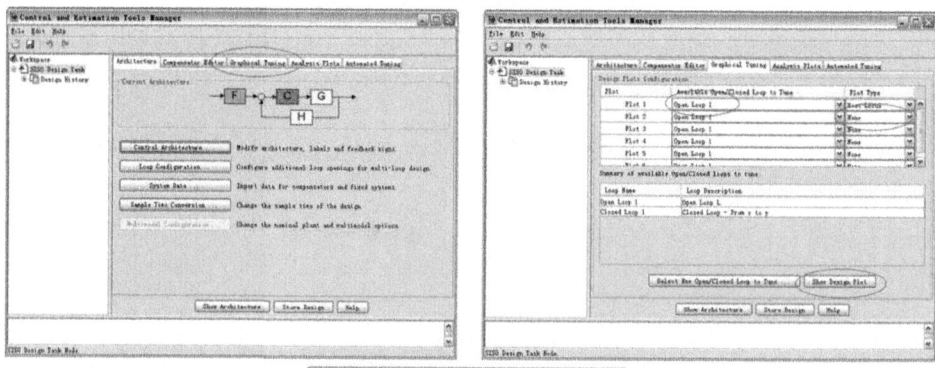

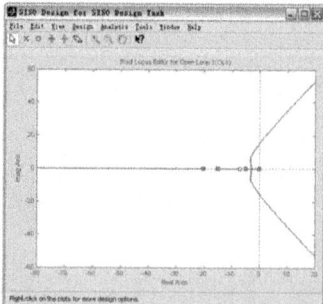

In the same fashion, select the tab labeled Analysis Plots. Within this window, for Plot 1, select Step. In the Contents of Plotssubwindow, select Closed Loop r to y for Plot 1. If the window does not automatically pop up, click the button labeled Show Analysis Plot.

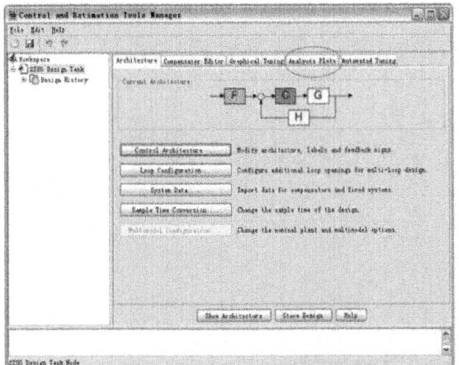
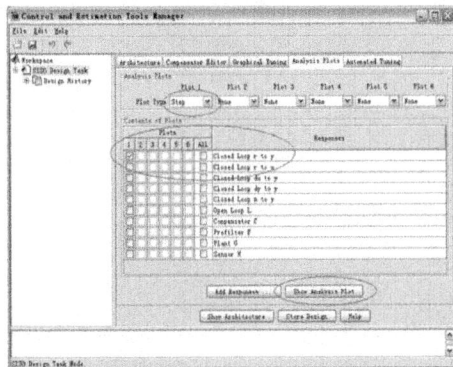

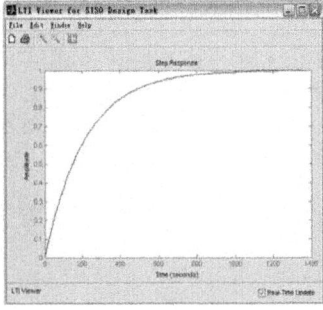

The next thing to do is to add the design requirements to the Root Locus plot. This is done directly on the plot by right-clicking and selecting Design Requirements, New. Design requirements can be set for the Settling Time, the Percent Overshoot, the Damping Ratio, the Natural Frequency, or a Region Constraint. There is no direct requirement for Rise Time, but the natural frequency can be used for this.

Here, we will set the design requirements for the damping ratio and the natural frequency just like was done with sgrid. Recall that the requirements call for $\zeta = 0.7$ and $\omega_n = 1.8$. Set these within the design requirements. On the plot, any area which is still white, is an acceptable region for the poles.

Zoom into the Root Locus by right-clicking on the axis and select Properties, then click the label Limits. Change the real axis to −25 to 5 and the imaginary to −2.5 to 2.5.

Also, we can see the current values of some key parameters in the response. In the Step response, right-click on the plot and go to Characteristics and select Peak Response. Do the same for the Rise Time. There should now be two large dots on the screen indicating the location of these parameters.

Both plots should appear as shown here:

As the characteristics show on the Step response, the overshoot is acceptable, but the rise time is incredibly off.

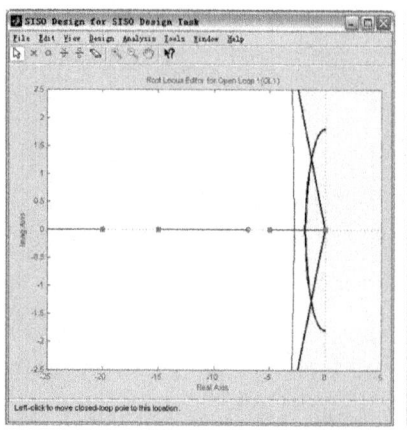 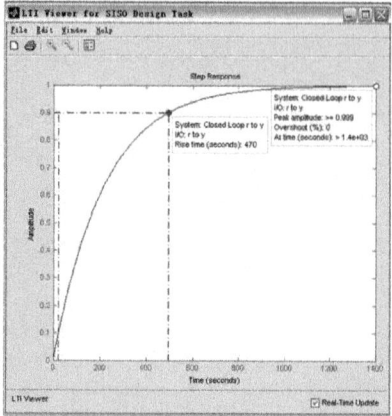

To fix this, we need to choose a new value for the gain K. Similarly to the `rlocfind` command, the gain of the controller can be changed directly on the root locus plot. Click and drag the pink box on the origin to the acceptable area where the poles have an imaginary component as shown below.

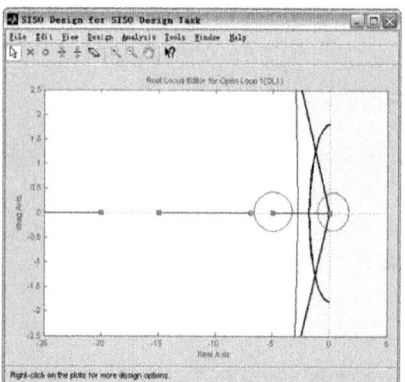 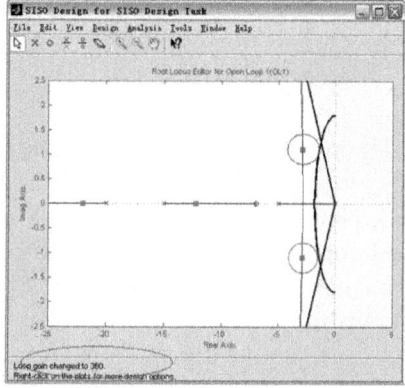

At the bottom of the plot, it can be seen that the loop gain has been changed to 361. Looking at the Step response, both of the values are acceptable for our requirements.

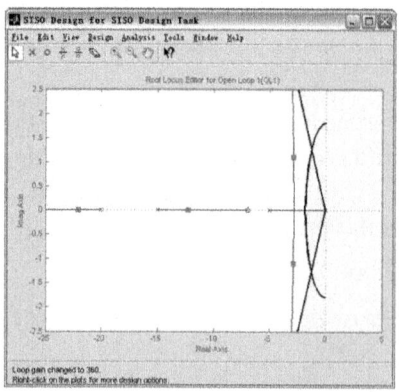 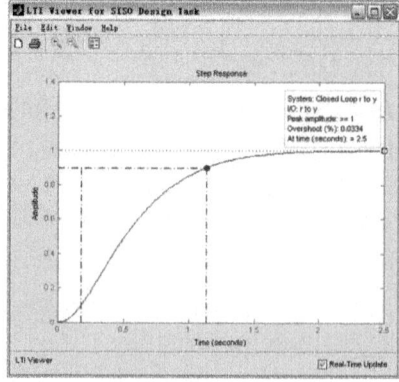

PID CONTROLLER FEATURES

In order for any PID controller to be practical, it must be able to do more than just implement the PID equation. This chapter identifies and explains some of the basic features found on most (but not all!) modern PID controllers:

- Manual versus Automatic mode
- Output tracking
- Setpoint tracking
- Alarming
- PV characterization and damping
- Setpoint limits
- Output limits
- PID tuning security

Manual and Automatic Modes

When a controller continually calculates output values based on PV and SP values over time, it is said to be operating inautomatic mode. This mode, of course, is what is necessary to regulate any process. There are times, however, when it is desirable to allow a human operator to manually "override" the automatic action of the PID controller. Applicable instances include process startup and shut-down events, emergencies, and maintenance procedures. A controller that is being "overridden" by a human being is said to be in manual mode.

A very common application of manual mode is during maintenance of the sensing element or transmitter. If an instrument technician needs to disconnect a process transmitter for calibration or replacement, the controller receiving that transmitter's signal cannot be left in automatic mode. If it is, then the controller may take sudden corrective action the moment the transmitter's signal goes dead. If the controller is first placed in manual mode before the technician disconnects the transmitter, however, the controller will ignore any changes in the PV signal, letting its output signal be adjusted at will by the human operator. If there is another indicator of the same process variable as the one formerly reported by the disconnected transmitter, the human operator may elect to read that other indicator and play the part of a PID controller, manually adjusting the final control element to maintain the alternate indicator at setpoint while the technician completes the transmitter's maintenance.

An extension of this "mode" concept applies to controllers configured to receive a setpoint from another device (called aremote or cascaded setpoint). In addition to an automatic and a manual mode selection, a third selection called cascadeexists to switch the controller's setpoint from human operator control to remote (or "cascade") control.

Output and Setpoint Tracking

The provision of manual and automatic operating modes creates a set of potential problems for the PID controller. If, for example, a PID controller is switched from automatic to manual mode by a human operator, and then the output is manually adjusted to some new value, what will the output value do when the controller is switched back to automatic mode? In some crude PID controller designs, the result would be an immediate "jump" back to the output value calculated by the PID equation while the controller was in manual. In other words, some controllers never stop evaluating the PID equation – even while in manual mode – and will default to that automatically-calculated output value when the operating mode is switched from manual to automatic.

This can be very frustrating to the human operator, who may wish to use the controller's manual mode as a way to change the controller's bias value. Imagine, for example, that a PD controller (no integral action) is operating in automatic mode at some low output value, which happens to be too low to achieve the desired setpoint. The operator switches the controller to manual mode and then raises the output value, allowing the process variable to approach setpoint. When PV nearly equals SP, the operator switches the controller's mode back to automatic, expecting the PID equation to start working again from this new starting point. In a crude controller, however, the output would jump back to some lower value, right where the PD equation would have placed it for these PV and SP conditions.

A feature designed to overcome this problem – which is so convenient that I consider it an essential feature of any controller with a manual mode – is called output tracking. With output tracking, the bias value of the controller shifts every time the controller is placed into manual mode and the output value manually changed. Thus, when the controller is switched from manual mode to automatic mode, the output does not immediately jump to some previously-calculated value, but rather "picks up" from the last manually-set value and begins to control from that point as dictated by the PID equation. In other words, output tracking allows a human operator to arbitrarily offset the output of a PID controller by switching to manual mode, adjusting the output value, and then switching back to automatic mode. The output will continue its automatic action from this new starting point instead of the old starting point.

A very important application of output tracking is in the manual correction of integral wind-up (sometimes called reset windup or just windup). This is what happens to a controller with integral action if for some reason the process variablecannot achieve setpoint no matter how far the output signal value is driven by integral action. An example might be on a temperature controller where the source of heat for the process is a steam system. If the steam system shuts down, the temperature controller cannot warm the process up to the temperature setpoint value no matter how far open the steam valve is driven by integral action. If the steam system is shut down for too long, the result will be a controller output saturated at maximum value in a futile attempt to warm the process.

If and when the steam system starts back up, the controller's saturated output will now send too much heating steam to the process, causing the process temperature to overshoot setpoint until integral action drives the output signal back down to some reasonable level. This situation may be averted, however, if the operator switches the temperature controller to manual mode as soon as the steam system shuts down. Even if this preventative step is not taken, the problem of overshoot may be averted upon steam system start-up if the operator uses output tracking by quickly switching the controller into manual mode, adjusting the output down to a reasonable level, and then switching back into automatic mode so that the controller's output value is no longer "wound up" at a high level.

A similar feature to output tracking – also designed for the convenience of a human operator switching a PID controller between automatic and manual modes – is called setpoint tracking. The purpose of setpoint tracking is to equalize SP and PV while the controller is in manual mode, so that when the controller gets switched back into automatic mode, it will begin its automatic operation with zero error (PV = SP).

This feature is most useful during system start-ups, where the controller may have difficulty Controlling the process in automatic mode under unusual conditions. Operators often prefer to run certain control loops in manual mode from the time of initial start-up until such time that the process is near normal operating conditions. At that point, when the operator is content with the stability of the process, the controller is assigned the responsibility of maintaining the process at setpoint. With setpoint tracking present in the controller, the controller's SP value will be held equal to the PV value (whatever that value happens to be) for the entire time the controller is in manual mode. Once the operator decides it is proper to switch the controller into automatic mode, the SP value freezes at that last manual-mode PV value, and the controller will continue to control the PV at that SP value. Of course, the operator is free to adjust the SP value to any new value while the controller is in automatic mode, but this is at the operator's discretion.

Without setpoint tracking, the operator would have to make a setpoint adjustment either before or after switching the controller from manual mode to automatic mode, in order to ensure the controller was properly set up to maintain the process variable at the desired value. With setpoint tracking, the setpoint value will default to the process variable value when the controller was last in manual mode, which (it is assumed) will be close enough to the desired value to suffice for continued operation.

Unlike output tracking, for which there is virtually no reason not to have the feature present in a PID controller, there may very well be applications where we do not wish to have setpoint tracking. For some processes, the setpoint value should remain fixed at all times, and as such it would be undesirable to have the setpoint value drift around with the process variable value every time the controller was placed into manual mode.

Alarm capabilities

A common feature on many instrument systems is the ability to alert personnel to the onset of abnormal process conditions. The general term for this function is alarm. Process alarms may be triggered by process switches directly sensing abnormal conditions (*e.g.* high-temperature switches, low-level alarms, low-flow alarms, *etc.*), in which case they are called hard alarms. A soft alarm, by contrast, is an alarm triggered by some continuous measurement (*i.e.* a signal from a process transmitter rather than a process switch) exceeding a pre-programmed alarm limit value.

Since PID controllers are designed to input continuous process measurements, it makes sense that a controller could be equipped with programmable alarm limit values as well, to provide "soft" alarm capability without adding additional instruments to the loop. Not only is PV alarming easy to implement in most PID controllers, but deviation alarming is easy to implement as well. A "deviation alarm" is a soft alarm triggered by excessive deviation (error) between PV and SP. Such an event indicates control problems, since a properly-operating feedback loop should be able to maintain reasonable agreement between PV and SP at all times.

Alarm capabilities find their highest level of refinement in modern distributed control systems (DCS), where the networked digital controllers of a DCS provide convenient access and advanced management of hard and soft alarms alike. Not only can alarms be accessed from virtually any location in a facility in a DCS, but they are usually time-stamped and archived for later analysis, which is an extremely important feature for the analysis of emergency events, and the continual improvement of process safety.

Output and Setpoint Limiting

In some process applications, it may not be desirable to allow the controller to automatically manipulate the final control element (control valve, variable-speed motor, heater) over its full 0% -100% range. In such applications, a useful controller feature is an output limit. For example, a PID flow controller may be configured to have a minimum output limit of 5%, so that it is not able to close the control valve any further than the 5% open position in order to maintain "minimum flow" through a pump. The valve may still be fully closed (0% stem position) in manual mode, but just not in automatic mode.

Similarly, setpoint values may be internally limited in some PID controllers, such that an operator cannot adjust the setpoint above some limiting value or below some other limiting value. In the event that the process variable must be driven outside these limits, the controller may be placed in manual mode and the process "manually" guided to the desired state by an operator.

Security

There is justifiable reason to prevent certain personnel from having access to certain parameters and configurations on PID controllers. Certainly, operations personnel need access to setpoint adjustments and automatic/manual mode controls, but it may be unwise to grant those same operators unlimited access to PID tuning constants and output limits. Similarly, instrument technicians may require access to a PID controller's tuning parameters, but perhaps should be restricted from editing configuration programs maintained by the engineering staff.

Most digital PID controllers have some form of security access control, allowing for different levels of permission in altering PID controller parameters and configurations. Security may be crude (a hidden switch located on a printed circuit board, which only the maintenance personnel should know about), sophisticated (login names and passwords, like a multi-user computer system), or anything in between, depending on the level of development invested in the feature by the controller's manufacturer.

An interesting solution to the problem of security in the days of analog control systems was the architecture of Foxboro's SPEC 200 analog electronic control system. The controller displays, setpoint adjustments, and auto/manual mode controls were located on the control room panel where anyone could access them. All other adjustments (PID settings, alarm settings, limit settings) could be located in the nest area where all the analog circuit control cards resided. Since the "nest" racks could be physically located in a room separate from the control room, personnel access to the nest room served as access security to these system parameters.

At first, the concept of controller parameter security may seen distrustful and perhaps even insulting to those denied access, especially when the denied persons possess the necessary knowledge to understand the functions and consequences of those parameters. It is not uncommon for soft alarm values to be "locked out" from operator access despite the fact that operators understand very well the purpose and functions of these alarms. At some facilities, PID tuning is the exclusive domain of process engineers, with instrument technicians and operators alike barred from altering PID tuning constants even though some operators and many technicians may well understand PID controller tuning.

When considering security access, there is more to regard than just knowledge or ability. At a fundamental level, security is a task of limiting access commensurate with responsibility. In other words, security restrictions exist to exclude those not charged with particular responsibilities. Knowledge and ability are necessary conditions of responsibility (*i.e.* one cannot reasonably be held responsible for something beyond their knowledge or control), but they are not sufficientconditions of responsibility (*i.e.* knowing how to, and being able to perform a task does not confer responsibility for that task getting completed). An operator may very well understand how and why a soft alarm on a controller works, but the responsibility for altering the alarm value may reside with someone else whose job description it is to ensure the alarm values correspond to plant-wide policies.

Chapter 5

PID CONTROLLER ALGORITHMS AND OPTIONS

PIDTUNE

PID tuning algorithm for linear plant model

Syntax

```
C = pidtune(sys,type)
C = pidtune(sys,C0)
C = pidtune(sys,type,wc)
C = pidtune(sys,C0,wc)
C = pidtune(sys,...,opts)
[C,info] = pidtune(...)
```

Description

    C = pidtune(sys,type) designs a PID controller of type type for the plant sys. If type specifies a one-degree-of-freedom (1-DOF) PID controller, then the controller is designed for the unit feedback loop as illustrated:

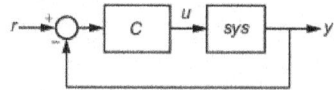

    If type specifies a two-degree-of-freedom (2-DOF) PID controller, then pidtune designs a 2-DOF controller as in the feedback loop of this illustration:

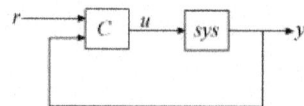

pidtune tunes the parameters of the PID controller C to balance performance (response time) and robustness (stability margins).

C = pidtune(sys,C0) designs a controller of the same type and form as the controller C0. If sys and C0 are discrete-time models, C has the same discrete integrator formulas as C0.

C = pidtune(sys,type,wc) and C = pidtune(sys,C0,wc) specify a target value wc for the first 0 dB gain crossover frequency of the open-loop response.

C = pidtune(sys,...,opts) uses additional tuning options, such as the target phase margin. Use pidtuneOptions to specify the option set opts.

[C,info] = pidtune(...) returns the data structure info, which contains information about closed-loop stability, the selected open-loop gain crossover frequency, and the actual phase margin.

Input Arguments

sys	Single-input, single-output dynamic system model of the plant for controller design. sys can be: • Any type of SISO dynamic system model, including Numeric LTI models and identified models. If sys is a tunable or uncertain model, pidtune designs a controller for the current or nominal value of sys. • A continuous- or discrete-time model. • Stable, unstable, or integrating. A plant with unstable poles, however, might not be stabilizable under PID control. • A model that includes any type of time delay. A plant with long time delays, however, might not achieve adequate performance under PID control. • An array of plant models. If sys is an array, pidtune designs a separate controller for each plant in the array. If the plant has unstable poles, and sys is one of the following: • A frd model • A ss model with internal time delays that cannot be converted to I/O delays you must use pidtuneOptions to specify the number of unstable poles in the plant, if any.

type	Controller type of the controller to design, specified as a string. The term *controller type* refers to which terms are present in the controller action. For example, a PI controller has only a proportional and an integral term, while a PIDF controller contains proportional, integrator, and filtered derivative terms. The type string can take the values summarized below. 1-DOF Controllers • 'P' — Proportional only • 'I' — Integral only • 'PI' — Proportional and integral • 'PD' — Proportional and derivative • 'PDF' — Proportional and derivative with first-order filter on derivative term • 'PID' — Proportional, integral, and derivative • 'PIDF' — Proportional, integral, and derivative with first-order filter on derivative term 2-DOF Controllers • 'PI2' — 2-DOF proportional and integral • 'PD2' — 2-DOF proportional and derivative • 'PDF2' — 2-DOF proportional and derivative with first-order filter on derivative term • 'PID2' — 2-DOF proportional, integral, and derivative • 'PIDF2' — 2-DOF proportional, integral, and derivative with first-order filter on derivative term 2-DOF Controllers with Fixed Setpoint Weights • 'I-PD' — 2-DOF PID with $b = 0, c = 0$ • 'I-PDF' — 2-DOF PIDF with $b = 0, c = 0$ • 'ID-P' — 2-DOF PID with $b = 0, c = 1$ • 'IDF-P' — 2-DOF PIDF with $b = 0, c = 1$ • 'PI-D' — 2-DOF PID with $b = 1, c = 0$ • 'PI-DF' — 2-DOF PIDF with $b = 1, c = 0$ Controller Form When you use the type input, pidtune designs a controller in parallel (pid or pid2) form. Use the input C0 instead of type if you want to design a controller in standard (pidstd or pidstd2) form. If sys is a discrete-time model with sample time Ts, pidtune designs a discrete-time controller with the same Ts. The controller has the ForwardEuler discrete integrator formula for both integral and derivative actions. Use the input C0 instead of type if you want to design a controller having a different discrete integrator formula.

CO	PID controller setting properties of the designed controller, specified as a pid, pidstd, pid2, orpidstd2 object. If you provide C0, pidtune:
	• Designs a controller of the type represented by C0.
	• Returns a pid controller, if C0 is a pid controller.
	• Returns a pidstd controller, if C0 is a pidstd controller.
	• Returns a 2-DOF pid2 controller, if C0 is a pid2 controller.
	• Returns a 2-DOF pidstd2 controller, if C0 is a pidstd2 controller.
	• Returns a controller with the same Iformula and Dformula values as C0, if sys is a discrete-time system.
wc	Target value for the 0 dB gain crossover frequency of the tuned open-loop response. Specify wc in units of radians/TimeUnit, where TimeUnit is the time unit of sys. The crossover frequency wcroughly sets the control bandwidth. The closed-loop response time is approximately 1/wc.
	Increase wc to speed up the response. Decrease wc to improve stability. When you omit wc,pidtune automatically chooses a value, based on the plant dynamics, that achieves a balance between response and stability.
opts	Option set specifying additional tuning options for the pidtune design algorithm, such as target phase margin or design focus. Use pidtuneOptions to create opts.

Output Arguments

C	Controller designed for sys. If sys is an array of linear models, pidtune designs a controller for each linear model and returns an array of PID controllers.
	Controller form:
	• If the second argument to pidtune is type, C is a pid or pid2 controller.
	• If the second argument to pidtune is C0:
	o C is a pid controller, if C0 is a pid object.
	o C is a pidstd controller, if C0 is a pidstd object.
	o C is a pid2 controller, if C0 is a pid2 object.
	o C is a pidstd2 controller, if C0 is a pidstd2 object.
	Controller type:
	• If the second argument to pidtune is type, C generally has the specified type.
	• If the second argument to pidtune is C0, C generally has the same type as C0.
	In either case, however, where the algorithm can achieve adequate performance and robustness using a lower-order controller than specified with type or C0, pidtune returns a C having fewer actions than specified. For example, C can be a PI controller even though type is 'PIDF'.
	Time domain:
	• C has the same time domain as sys.
	• If sys is a discrete-time model, C has the same sample time as sys.
	• If you specify C0, C has the same Iformula and Dformula as C0. If no C0 is specified, bothIformula and Dformula are Forward Euler.
	If you specify C0, C also obtains model properties such as InputName and Output-Name from C0.
info	Data structure containing information about performance and robustness of the tuned PID loop. The fields of info are:
	• Stable — Boolean value indicating closed-loop stability. Stable is 1 if the closed loop is stable, and 0 otherwise.
	• CrossoverFrequency — First 0 dB crossover frequency of the open-loop system C*sys, inrad/TimeUnit, where TimeUnit is the time units specified in the TimeUnit property of sys.
	• PhaseMargin — Phase margin of the tuned PID loop, in degrees.
	If sys is an array of plant models, info is an array of data structures containing information about each tuned PID loop.

PID Controller Design at the Command Line

This example shows how to design a PID controller for the plant given by:

$$sys = \frac{1}{(s+1)^3}.$$

As a first pass, create a model of the plant and design a simple PI controller for it.

```
sys = zpk([],[-1 -1 -1],1);
[C_pi,info] = pidtune(sys,'PI')
C_pi =

            1
Kp + Ki * ---

            s
with Kp = 1.14, Ki = 0.454
Continuous-time PI controller in parallel form.
info =
              Stable: 1
   CrossoverFrequency: 0.5205
          PhaseMargin: 60.0000
```

C_pi is a pid controller object that represents a PI controller. The fields of info show that the tuning algorithm chooses an open-loop crossover frequency of about 0.52 rad/s.

Examine the closed-loop step response (reference tracking) of the controlled system.

```
T_pi = feedback(C_pi*sys, 1);
step(T_pi)
```

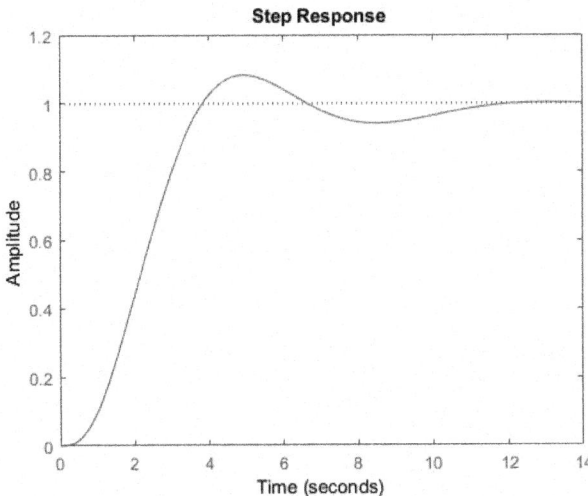

To improve the response time, you can set a higher target crossover frequency than the result that pidtune automatically selects, 0.52. Increase the crossover frequency to 1.0.

```
[C_pi_fast,info] = pidtune(sys,'PI',1.0)
C_pi_fast =

                1
  Kp + Ki * ---

                s
  with Kp = 2.83, Ki = 0.0495
Continuous-time PI controller in parallel form.
info =


              Stable: 1
  CrossoverFrequency: 1
         PhaseMargin: 43.9973
```

The new controller achieves the higher crossover frequency, but at the cost of a reduced phase margin.

Compare the closed-loop step response with the two controllers.

```
T_pi_fast = feedback(C_pi_fast*sys,1);
step(T_pi,T_pi_fast)
axis([0 30 0 1.4])
legend('PI','PI,fast')
```

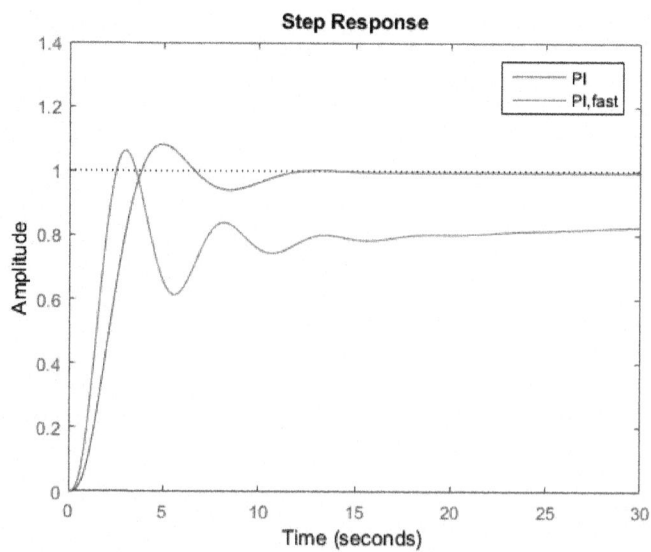

This reduction in performance results because the PI controller does not have enough degrees of freedom to achieve a good phase margin at a crossover frequency of 1.0 rad/s. Adding a derivative action improves the response.

Design a PIDF controller for Gc with the target crossover frequency of 1.0 rad/s.

```
[C_pidf_fast,info] = pidtune(sys,'PIDF',1.0)
C_pidf_fast =

                1              s
  Kp + Ki * --- + Kd * --------
                s            Tf*s+1

  with Kp = 2.72, Ki = 0.985, Kd = 1.72, Tf = 0.00875
Continuous-time PIDF controller in parallel form.
info =
                Stable: 1
    CrossoverFrequency: 1
        PhaseMargin: 60.0000
```

The fields of info show that the derivative action in the controller allows the tuning algorithm to design a more aggressive controller that achieves the target crossover frequency with a good phase margin.

Compare the closed-loop step response and disturbance rejection for the fast PI and PIDF controllers.

```
T_pidf_fast = feedback(C_pidf_fast*sys,1);
step(T_pi_fast, T_pidf_fast);
axis([0 30 0 1.4]);
legend('PI,fast','PIDF,fast');
```

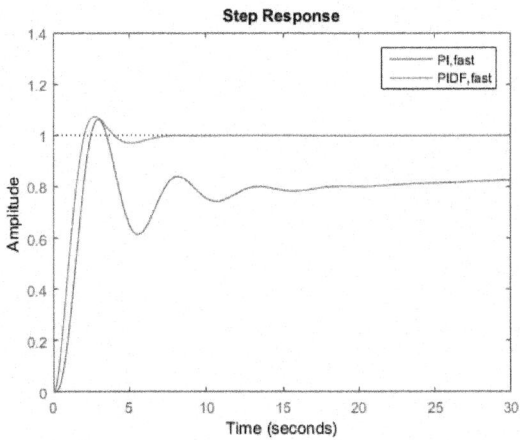

You can compare the input (load) disturbance rejection of the controlled system with the fast PI and PIDF controllers. To do so, plot the response of the closed-loop transfer function from the plant input to the plant output.

```
S_pi_fast = feedback(sys,C_pi_fast);
S_pidf_fast = feedback(sys,C_pidf_fast);
step(S_pi_fast,S_pidf_fast);
axis([0 50 0 0.4]);
legend('PI,fast','PIDF,fast');
```

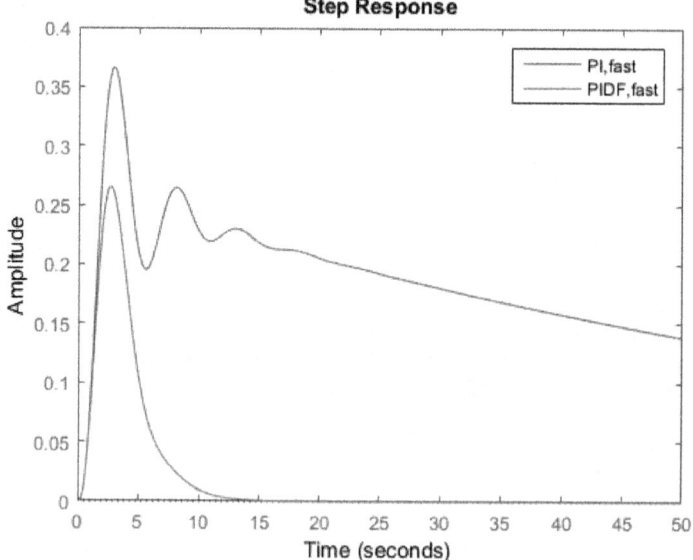

This plot shows that the PIDF controller also provides faster disturbance rejection.

Design Standard-Form PID Controller

Design a PID controller in standard form for the plant defined by
$$sys = \frac{(\quad)3}{1\ s+1}.$$
To design a controller in standard form, use a standard-form controller as the C0 argument to pidtune.

```
sys = zpk([],[-1 -1 -1],1);
C0 = pidstd(1,1,1);
C = pidtune(sys,C0)
C =
```

```
            1       1
  Kp * (1 + ---- * --- + Td * s)
            Ti      s


  with Kp = 2.18, Ti = 2.36, Td = 0.591


  Continuous-time PID controller in standard form
```

Specify Integrator Discretization Method

Design a discrete-time PI controller using a specified method to discretize the integrator.

If your plant is in discrete time, pidtune automatically returns a discrete-time controller using the default Forward Euler integration method. To specify a different integration method, use pid or pidstd to create a discrete-time controller having the desired integration method.

```
sys = c2d(tf([1 1],[1 5 6]),0.1);
C0 = pid(1,1,'Ts',0.1,'IFormula','BackwardEuler');
C = pidtune(sys,C0)
C =

            Ts*z
  Kp + Ki * ------
            z-1


  with Kp = -0.518, Ki = 10.4, Ts = 0.1


  Sample time: 0.1 seconds
  Discrete-time PI controller in parallel form.
```

Using C0 as an input causes pidtune to design a controller C of the same form, type, and discretization method as C0. The display shows that the integral term of C uses the Backward Euler integration method.

Specify a Trapezoidal integrator and compare the resulting controller.

```
C0_tr = pid(1,1,'Ts',0.1,'IFormula','Trapezoidal');
Ctr = pidtune(sys,C_tr)
Ctr =
```

```
        Ts*(z+1)
  Ki * --------
        2*(z-1)


  with Ki = 10.4, Ts = 0.1


  Sample time: 0.1 seconds
  Discrete-time I-only controller.
```

Design 2-DOF PID Controller

Design a 2-DOF PID Controller for the plant given by the transfer function:

$$G(s) = \frac{1}{s^2 + 0.5s + 0.1}.$$

Use a target bandwidth of 1.5 rad/s.

```
wc = 1.5;
G = tf(1,[1 0.5 0.1]);
C2 = pidtune(G,'PID2',wc)

C2 =

                  1
  u = Kp (b*r-y) + Ki --- (r-y) + Kd*s (c*r-y)
                  s


  with Kp = 1.26, Ki = 0.255, Kd = 1.38, b = 0.665, c = 0


  Continuous-time 2-DOF PID controller in parallel form.
```

Using the type string 'PID2' causes pidtune to generate a 2-DOF controller, represented as a pid2 object. The display confirms this result. The display also shows that pidtune tunes all controller coefficients, including the setpoint weights b and c, to balance performance and robustness.

PID CONTROLLER ALGORITHMS

Controller manufacturers arrange the Proportional, Integral and Derivative modes into three different controller algorithms or controller structures. These are called Interactive, Noninteractive, and Parallel algorithms. Some controller manufacturers allow you to choose between different controller algorithms as a configuration option in the controller software.

Interactive Algorithm

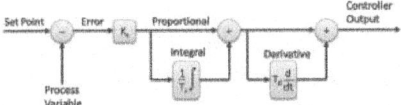

Fig. : Interactive Controller Algorithm

$$CO = K_c \left[E + \frac{1}{T_i} \int E \cdot dt \right] \times \left[1 + T_d \frac{d \cdot}{dt} \right]$$

The oldest controller algorithm is called the Series, Classical, Real or Interactive algorithm. The original pneumatic and electronic controllers had this algorithm and it is still found it in many controllers today. The Ziegler-Nichols PID tuning rules were developed for this controller algorithm.

Noninteractive Algorithm

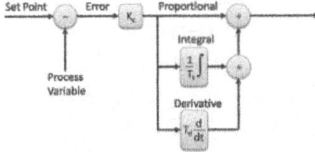

Fig. : Noninteractive Controller Algorithm.

$$CO = K_c \left[E + \frac{1}{T_i} \int E \cdot dt + T_d \frac{dE}{dt} \right]$$

The Noninteractive algorithm is also called the Ideal, Standard or ISA algorithm. The Cohen-Coon and Lambda PID tuning rules were designed for this algorithm.

Note: If no derivative is used (*i.e.* $T_d = 0$), the interactive and noninteractive controller algorithms are identical.

Parallel Algorithm

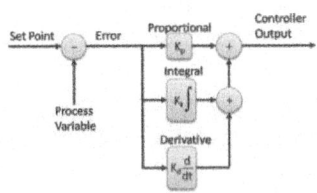

Fig. : Parallel Controller Algorithm.

$$CO = K_p \times E + K_i \int E \cdot dt + K_d \frac{dE}{dt}$$

Some academic textbooks discuss the parallel form of PID controller, but it is also used in some DCSs and PLCs. This algorithm is simple to understand, but not intuitive to tune. The reason is that it has no controller gain (affecting all three control modes), it has a proportional gain instead (affecting only the proportional mode). Adjusting the proportional gain should be supplemented by adjusting the integral and derivative settings at the same time. Try to not use this controller algorithm if possible (in some DCSs it is an option, so select the alternative).

Significance of Different Algorithms

The biggest difference between the controller algorithms is that the Parallel controller has a true Proportional Gain (Kp), while the other two algorithms have a Controller Gain (Kc). Controller Gain affects all three modes (Proportional, Integral and Derivative) of the Series and Ideal controllers, while Proportional Gain affects only the Proportional mode of a Parallel controller.

This difference has a major impact on the tuning of the controllers. All the popular tuning rules (Ziegler-Nichols, Cohen-Coon, Lambda, and others) assume the controller does not have a parallel structure and therefore has a Controller Gain. To tune a Parallel controller using any of these rules, the Integral time has to be divided and derivative time multiplied by the calculated Controller Gain.

The second difference between the controller algorithms is the interaction between the Integral and Derivative modes of the Series (Interactive) controller. This, of course, is only of significance if the Derivative mode is used. In most PID controller applications, Derivative mode is not used. Formulas have been developed for converting tuning settings between Ideal and Series controller algorithms.

Units of Measure of Tuning Settings

Another very important difference between controllers lies in the units of measure of the tuning settings. There are three differences.

1. Most controller types (*e.g.* Honeywell Experion, Emerson DeltaV, ABB Bailey) use Controller Gain, while some (*e.g.* Foxboro I/A, Yokogawa CS3000) use Proportional Band (PB). The conversion between the two is easy once you know which one is being used: PB = 100% / Kc.

2. Many controllers (*e.g.* Siemens APACS) use minutes as the unit for Integral and Derivative modes, but some controllers (*e.g.* Emerson DeltaV) use seconds.

3. Some controllers (*e.g.* ABB Mod 300) use Time for their Integral unit, while others (*e.g.* Allen-Bradley SLC500) use Repeats/Time. These are reciprocals of each other.

The first controller I ever tried to tune used Proportional Band, but at the time, I had never heard of this concept. Needless to say, when I entered my calculated Kc of 1.2 into its PB setting, the loop became wildly unstable. It did not take me long to realize that I should read up on PID controllers before trying to tune one again.

Other Differences

Beyond the differences mentioned above, controllers also differ in the way the changes on controller output is calculated (positional and velocity algorithms), in the way Proportional and Derivative modes act on set point changes, in the way the Derivative mode is limited/filtered, as well as a interesting array of other minor differences. These differences are normally subtle, and should not affect your tuning.

When tuning controllers, always find out what structure the controller has and what units it is using.

THE PID ALGORITHM

There is no single PID algorithm. Different fields using feedback control have probably used different algorithms ever since math was introduced to feedback control. This Web page (a single file, of four pages, no pictures) is written for people in the process industries, for that is the only field in which I (David W. St. Clair) have experience. Even in that single field, which has been served by companies such as ABB (formerly Taylor), Bailey, Fisher, Foxboro, Honeywell, Moore Products, Yokogawa and others there is no standard algorithm. Perhaps years ago there was (or for most practical purposes was), but today there are many algorithms. Also there is no standard terminology. For the person interested in tuning controllers for the process industries it has become a bit more complicated, because the rules and procedures you would use to tune with one algorithm are not the ones you would use to tune with another. Also, with the added features available with computers, some of the configurations can become quite complex. This page does not begin to address those, but you certainly need to understand what your basic building block is.

The purpose of this Web page is to focus on the fact that there are differences and to describe them (or at least to alert you to look for them). No reference to the algorithm of specific manufacturers is given. If you are tuning controllers you must know the algorithm of the equipment you are using. For that you should read the information provided by the manufacturer. Even the words used to identify an algorithm are ambiguous. You should look at the equation. This is unfortunate because many persons assigned the responsibility of tuning process industry controllers are not comfortable with equations. If you are reading this as preparation for writing a PID algorithm, it will alert you to the fact that there is more to it than you might have thought. Indeed, the feedback I have from knowledgeable people is that even the experts can slip up.

Presently there are three basic forms of the PID algorithm. These will be discussed in turn. After that there is a short discussion of other aspects of any algorithm which must be considered to write the digital program for one.

Expressed by their Laplace transforms the three forms are:

First form: $Kc(1 + 1/Tis)(1 + Tds)/(1 + Tds/Kd)$

Second form: $Kc'(1 + 1/Ti's + Td's)$

Third form: $Kc'' + 1/Ti''s + Td''s$

where

Kc, Kc' and Kc'' relate to the P part of PID

Ti, Ti' and Ti'' relate to the I part of PID

Td, Td' and Td'' relate to the D part of PID

s is the Laplace notation for derivative with respect to time

Kd is the derivative gain

I have deliberately not assigned a name to any of these forms yet. Also I have not given a name to the variables. Both will come later as each algorithm. The second and third forms can be made equivalent to the first form (provided derivative is handled appropriately), but the first form cannot duplicate all combinations available in the second and third forms. The second and third forms can be made equal to each other. For most practical purposes one algorithm is not better than another, just different.

The First form of the PID Algorithm

This first form is called "series" or "interacting" or "analog" or "classical". The variables are:

Kc = controller gain = 100/proportional band

Ti = Integral or reset time = 1/reset rate in repeats/time

Td = derivative time

Kd = derivative gain

Early pneumatic controllers were probably designed more to meet mechanical and patent constraints than by a zeal to achieve a certain algorithm. Later pneumatic controllers tended to have an algorithm close to this first form. Electronic controllers of major vendors tended to use this algorithm. It is what process industry control users were used to at the time. If you are unsure what algorithm is being used for the controller you are tuning, find out what it is before you start to tune.

I did not follow closely the evolution of algorithms as digital controllers were introduced. It is my understanding that most major vendors of digital controllers provide this algorithm as basic, and many provide the second form as well. Also, many provide several variations (I'm told Allen-Bradley has 10, and that other manufacturers are adding variations continually).

The choice of the word interacting is interesting. At least one author says that it is interacting in the time domain and noninteracting in the frequency domain. Another author disagrees with this distinction.

Second form of the PID Algorithm

The second form of the algorithm is called "noninteracting, or "parallel" or "ideal" or "ISA". I understand one manufacturer refers to this as "interacting", which serves to illustrate that terms by themselves may not tell you what the algorithm is. This form is used in most textbooks, I understand. I think it is unfortunate that textbooks do not at least recognize the different forms. Most if not all books written for industry users rather than students recognize at least the first two forms. The basic difference between the first and second forms is in the way derivative is handled. If the derivative term is set to zero, then the two algorithms are identical. Since derivative is not used very often (and shouldn't be used very often) perhaps it is not important to focus on the difference. But it is important to anyone using derivative, and people who use derivative should know what they are doing. The parameters set in this form can be made equivalent (except for the treatment of gain-limiting on derivative) to those in the first form in this way:

$Kc' = ((Ti +Td)/Ti))Kc$, "effective" gain.

$Ti' = Ti + Td$, "effective" integral or reset time

$Td' = TiTd/(Ti + Td)$, "effective" derivative time

These conversions are made by equating the coefficients of s. Conversions in the reverse direction are:

$Kc = FKc'$

$Ti = FTi'$

$Td = Td'/F$

where

$F = 0.5 + sqrt(0.25 - Td'/Ti')$

Typically Ti is set about 4 to 8 times Td, so the conversion factor is not huge, but it is important to not loose sight of the correction. With this algorithm it is possible to have very troublesome combinations of Ti' and Td'. If Ti'<4Td' then the reset and derivative *times*, as differentiated from *settings*, become complex numbers, which can confuse tuning. Don't slip into these settings inadvertently! A very knowledgeable tuner may be able to take advantage of that characteristic in very special cases, but it is not for everyone, every day. Some companies advise to use the interacting form if available, simply to avoid that potential pitfall.

This algorithm also has no provision for limiting high frequency gain from derivative action, a virtually essential feature. In the first algorithm Kd is typically fixed at 10, or if adjustable, should typically be set somewhere in the range of 6 to 10. This desirable limiting of the derivative component is sometimes accomplished in this second form by writing it as:

$Kc'(1 + 1/Ti's + Td's)/(1 + Td's/Kd)$

or

$Kc'(1 + 1/Ti's + Td's/(1 + Td's/Kd))$

There are likely many variations on the theme.

The variables Kc', Ti' and Td' have been called "effective". In the Bode plot, IF Ti'>4Td', THEN Kc' is the minimum frequency-dependent gain (Kc is a frequency-independent gain). This is at a frequency which is midway between the "corners" defined by Ti and Td, which is also midway between the "effective " corners associated with Ti' and Td'. Ti' is always larger than Ti and Td' is always smaller than Td, which recognizes the slight spreading of the "effective" corners of the Bode plot as they approach each other.

This algorithm is also called the "ISA" algorithm. The ISA has no association with this algorithm. Apparently this attribution got started when someone working on the Fieldbus thought it would become "THE" algorithm. It didn't. Or hasn't. ANSI/ISA-S51.1-1979 is a standard on Process Instrumentation Terminology. While this is a standard on terminology, not algorithms, it uses the first form of the algorithm for examples and in its Bode plot for a PID controller. Another term used to identify this algorithm is "ideal". Think of this word as one to *identify* the algorithm, not *describe* it. It is true that it can do everything the first form can do, and more, provided the gain for derivative is handled appropriately. But settings which produce complex roots should be used only by the very knowledgeable.

Third form of the PID Algorithm

It is hard to know what to call this third form since it is so close to the second. It has been called "parallel", "ideal parallel", "noninteracting", "independent" and "gain independent". In one sense this third form is the second form rewritten. I understand this is the algorithm taught Electrical Engineers. The second and third forms can be made equal to each other by using the following substitutions:

$Kc'' = Kc'$

$Ti'' = Ti'/Kc'$

$Td'' = Kc'Td'$

They would only differ in what you call the tuning parameters. They are not gain, integral time and derivative time as those words are traditionally used in this field. Also, the option for limiting the gain from derivative action should be handled somehow, perhaps the same way as for form two. One option is as follows:

$Kc'' + 1/Ti''s + Td''s/(1+Td''s/K''d)$

The constraint in the second form that Ti'>4Td' to keep the roots real becomes Kc''Ti''>4Td''/Kc'', which is a bit more complicated.

Programming Considerations

There are many considerations in writing the program for a controller besides the decision on which basic algorithm to use. These include:

1. The option to have the derivative function act only on the process variable, not on set point changes.

2. The same option with regard to the proportional action. This option may be tied to the first, in that if you choose to have derivative act only on the process variable you get proportional action only on that also.

3. Provision for setpoint and process variable tracking, to permit bumpless automatic/manual transfers. You can have bumpless transfers without setpoint tracking. You can also transfer from manual to automatic without any bump due to proportional action. Aren't all these options wonderful!

4. Provision for reset windup protection.

5. Provision for a filter besides the one used to limit the derivative gain.

6. It is no simple matter to get digital derivative action to approach the quality of analog derivative action. No program can match it. This space is not intended to amplify on that problem, but simply to emphasize that it is a problem. It relates to sampling frequency and noise on the signal. Some algorithms use more than one back value of the controlled variable I believe. Also some manufacturers limit how low a derivative time may be set. It is very difficult for the user to know whether the derivative provided is doing a good job of achieving what could be achieved with derivative action.

7. Integral/reset action with digital controllers is not perfect. There is a phenomenon related to quantizing error, sampling time and long integral/reset times and calculating precision which prevents integrating to zero error. Apparently with more digits in the A/D converter and in the computer's math, this is becoming less and less of a problem.

8. There is the choice of having the algorithm be "velocity", sometimes called "incremental" (each calculation period a *change* in the output is calculated), or "position" (each calculation period the actual desired output is calculated). Apparently at one time there was a perception that the velocity algorithm did not have a reset windup problem, but this is not the case. The choice between the incremental and position algorithms seems to be a choice based on many considerations which are beyond the scope of this write-up.

9. There are options on filtering noise, such as providing a dead zone or a zone of low gain around the setpoint.

10. There are options to be considered in special cases, such as preventing reset windup in override and cascade situations.

11. Provision needs to be made for manual bias.

12. There must be other points to make to caution the novice. Does anyone want to suggest some?

THE CHAOS OF COMMERCIAL PID CONTROL

The design and tuning of a **three mode PID controller** follows the proven recipe we have used with success for P-Only control (*e.g.,* **here** and **here**) and PI Control (*e.g.,* **here,here** and **here**). The decisions and procedures we established for steps 1-3 of the **design and tuning recipe** in these previous studies remain unchanged as we move on to the PID algorithm.

Step 4 of the recipe remains the same as well. But it is essential in this step that we match the rules and correlations of step 4 with the particular controller algorithm form we are using.

The challenge arises because the number of PID algorithm forms available from hardware vendors increases markedly when derivative action is included. And unfortunately, these PID algorithms are **implemented in many different forms** across the commercial market.

The potential for confusion by even a careful practitioner is significant. For example:

- there are three popular PID algorithm forms, and
- each of these three forms have multiple parameters that are cast in different ways.

As a result, there are literally dozens of possible PID algorithm forms. Matching each controller form with its proper design rules and correlations requires careful attention if performed without the help of software tools.

Common Algorithm Forms

Listed below are the three common PID controller forms. If offered as an option by our vendor (most do offer it), **derivative on measured process variable (PV)** is the recommended PID form:

- Dependent, ideal PID controller form (derivative on measurement):

$$CO = CO_{bias} + Kc \cdot e(t) + \frac{Kc}{Ti} \int e(t)dt - Kc \cdot Td \frac{dPV}{dt}$$

- Dependent, interacting form (derivative on measurement):

$$CO = CO_{bias} + Kc\left(1 + \frac{Td}{Ti}\right)e(t) + \frac{Kc}{Ti} \int e(t)dt - KcTd \frac{dPV}{dt}$$

- Independent PID form (derivative on measurement):

$$CO = CO_{bias} + Kc \cdot e(t) + Ki \int e(t)dt - Kd \frac{dPV}{dt}$$

Where for the above:

CO = controller output signal **(the wire out)**

CO_{bias} = controller bias; set by **bumpless transfer**

$e(t)$ = current controller error, defined as SP – PV

SP = set point

PV = measured process variable **(the wire in)**

Kc = controller gain (also called proportional gain), a tuning parameter

Ki = integral gain, a tuning parameter

Kd = derivative gain, a tuning parameter

Ti = reset time, a tuning parameter

Td = derivative time, a tuning parameter

Tuning Parameters

Because there has been little standardization on nomenclature, the same tuning parameters can appear under different names in the commercial market. Perhaps more unfortunate, the same parameter can even have a different name within a single company's product line.

We will not attempt to list all of the different names here, though we will look at a solution to this issue later in this article. A few notes to consider:

1) The dependent forms appear most in products commonly used in the process industries, but the independent form is not uncommon.

2) The majority of DCS and PLC systems now use controller gain, Kc, for their dependent PID algorithms. There are notable exceptions, however, such as Foxboro who uses proportional band (PB = 100/Kc assuming PV and CO both range from 0 to 100%).

3) Reset time, Ti, is slightly more common for the dependent PID algorithms, though it is rarely called that in product documentation. Reset rate, defined as Tr = 1/Ti, comes in a close second. Again, the name for this parameter changes with product.

4) Most vendors use derivative time, Td, for their dependent PID algorithms, though few refer to it by that name in their product documentation.

Tune One, Tune Them All

Some good news in all this confusion is that the different forms, if tuned with the proper correlations, will perform exactly the same. No one form is better than another, it is just expressed differently.

In fact, we can show equivalence among the parameters, and thus algorithms, with these relations.

- *Proportional*

$$\text{Kc, ideal} = \text{Kc, interact}\left(1 + \frac{\text{Td, interact}}{\text{Ti, interact}}\right)$$

Kc, independent = Kc, ideal

- *Integral*

$$\text{Ti, ideal} = \text{Ti, interact}\left(1 + \frac{\text{Td, interact}}{\text{Ti, interact}}\right)$$

$$Ki, independent = \frac{Kc, ideal}{Ti, deal}$$

- *Derivative*

$$Td, ideal = \frac{Td, interact}{\left(1 + \dfrac{Td, interact}{Ti, interact}\right)}$$

Kd, independent = (Kc, ideal) (Td, ideal)

Though not presented here, analogous conversion relations can be developed for forms expressed using proportional band and/or reset rate.

Clarity in the Chaos

It is perhaps reasonable to hope that industrial practitioners will have an intuitive understanding of proportional, integral and derivative action. They might know the benefits each term offers and problems each presents. And experienced practitioners will know how design, tune and validate a PID implementation.

Expecting a practitioner to convert that knowledge and intuition over into the confusion of the commercial PID marketplace might not be so reasonable.

Given this, the best solution for those in the real world is to use software that lets us focus on the big picture while the software ensures that details are properly addressed.

Such productivity software should not only provide a "click and go" approach to algorithm and tuning parameter selection, but should also provide this information simply based on our choice of equipment manufacturer and product line.

For example, below is a portion of the controller manufacturer selection available in onecommercial software package:

If you select Allen Bradley, Emerson, and Honeywell in the above list, the choice of PID controllers for each company is shown in the next three images:

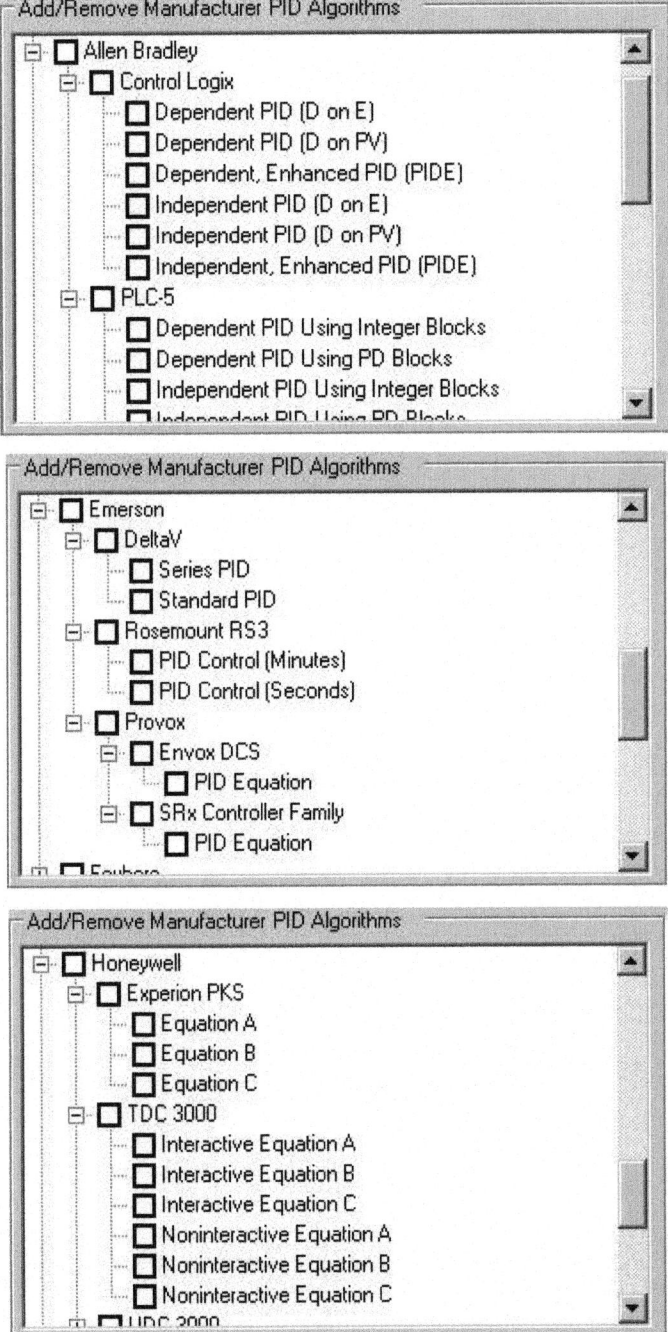

It is clear from these displays that there are different terms and many options for us to select from, all for PID control. And it may not be obvious that the different terms above refer to some version of our "basic three" PID forms.

Too much is at stake in a plant to ask a practitioner to keep track of it all. Software can get us past the details during PID controller design and tuning so we can focus on mission-critical control tasks like improving safety, performance and profitability.

Note: the Laplace domain is a subject that most control practitioners can avoid their entire careers, but it provides is a certain mathematical "elegance." Below, for example, are the three controller forms assuming derivative on error. Even without familiarity with Laplace, perhaps you will agree the three PID forms indeed look like part of the same family:

• *Dependent, Ideal (Non-interacting) Form*

$$\frac{CO(s)}{E(s)} = Kc\left(1 + \frac{1}{Ti\,s} + Td\,s\right)$$

• *Dependent, Interacting Form*

$$\frac{CO(s)}{E(s)} = Kc\left(1 + \frac{1}{Ti\,s}\right)(Td\,s + 1)$$

• *Independent Form*

$$\frac{CO(s)}{E(s)} = Kc + \frac{Ki}{s} + Kd\,s$$

THE NORMAL OR STANDARD PID ALGORITHM

The question arises quite often, "What is the normal or standard form of the PID (proportional-integral-derivative) algorithm?"

The answer is both simple and complex. Before we explore the answer, consider the screen displays shown below:

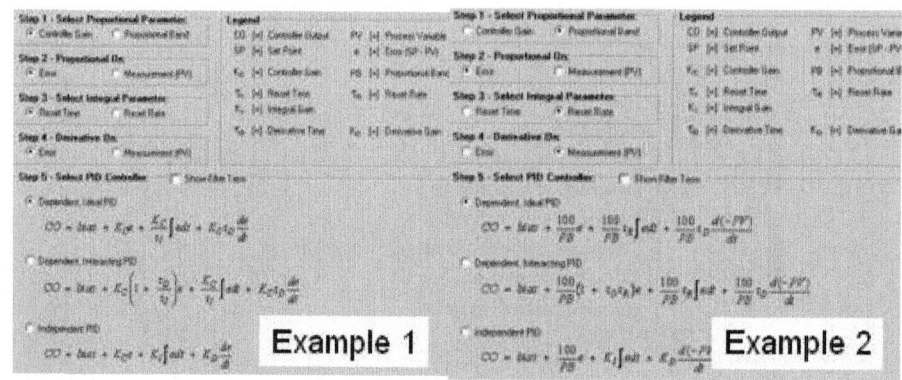

As shown in the screen displays:

There are three popular PID algorithm forms.

Each of the three algorithms has tuning parameters and algorithm variables that can be cast in different ways.

So your vendor might be using one of dozens of possible algorithm forms. And if you add a **filter term** to your controller, the number of possibilities increases substantially.

The Simple Answer

Any of the algorithms can deliver the same performance as any of the others. There is no control benefit from choosing one form over another. They are all standard or normal in that sense.

If you are considering a purchase, select the vendor that serves your needs the best and don't dwell on the specifics of the algorithm. Some things to consider include:

compatibility with existing controllers and associated hardware and software

cost

ease of installation and maintenance

reliability

your operating environment (is it clean? cool? dry?)

A More Complete Answer

Most of the different controller algorithm forms can be found in one vendor's product or another. Some vendors even use different forms within their own product lines.

And while the various forms are equally capable, each must be tuned (values for the adjustable parameters must be specified) using tuning correlations specifically designed for that particular control algorithm.

Commercial software makes it straightforward to get desired performance from any of them. But it is essential that you know your vendor and controller model number to ensure a correct match between controller form and computed tuning values.

The alternative to an orderly design methodology is a "guess and test" approach. While used by some practitioners, such trial and error tuning squanders valuable production time, consumes more feedstock and utilities than is necessary, generates additional waste and off-spec product, and can even present safety concerns.

We use some variation of the dependent, ideal PID controller form:

$$CO = CO_{bias} + Kc \cdot e(t) + \frac{Kc}{Ti} \int e(t)dt + Kc \cdot Td\frac{de(t)}{dt}$$

Where:

CO = controller output signal

CO_{bias} = controller bias

e(t) = current controller error, defined as SP – PV

SP = set point

PV = measured process variable

Kc = controller gain, a tuning parameter

Ti = reset time, a tuning parameter

Td = derivative time, a tuning parameter

To reinforce that the controllers all are equally capable, we occasionally use variations of the dependent, interacting form:

$$CO = CO_{bias} + Kc\left(1 + \frac{Td}{Ti}\right)e(t) + \frac{Kc}{Ti}\int e(t)dt + Kc \cdot Td\frac{de(t)}{dt}$$

or variations of the independent PID form:

$$CO = CO_{bias} + Kc \cdot e(t) + Ki\int e(t)dt + Kd\frac{de(t)}{dt}$$

Final Thoughts

Some of the subtle differences in algorithm form that we can exploit to improve control performance.

For example, derivative on error behaves different from **derivative on measured PV**. This is true for all of the algorithms. Derivative on error can "kick" after set point steps and this is rarely considered desirable behaviour. Thus, derivative on PV is recommended for industrial applications.

And if you are considering programming the controller yourself, it is not the algorithm form that is the challenge. The big hurdle is properly accounting for the **anti-reset windup and jacketing logic** to allow bumpless transition between operating modes.

PID CONTROLLER DESIGN

Key MATLAB commands used are: `sisotool`

From the main problem, the open-loop transfer function for the aircraft pitch dynamics is

$$P(s) = \frac{\Theta(s)}{\Delta(s)} = \frac{1.151s + 0.1774}{s^3 + 0739s^2 + 0.921s}$$

where the input is elevator deflection angle δ and the output is the aircraft pitch angle θ.

For the original problem setup and the derivation of the above transfer function please refer to the Aircraft Pitch: System Modeling page.

For a step reference of 0.2 radians, the design criteria are the following.

- Overshoot less than 10%
- Rise time less than 2 seconds
- Settling time less than 10 seconds
- Steady-state error less than 2%

Recall from the Introduction: PID Controller Design page that the transfer function for a PID controller is the following.

$$C(s) = K_p + \frac{K_i}{s} + K_d s \frac{K_d s^2 + K_p s + K_i}{s}$$

We will implement combinations of proportional (K_p), integral (K_i), and derivative (K_d) control in the unity feedback architecture shown below in order to achieve the desired system behaviour.

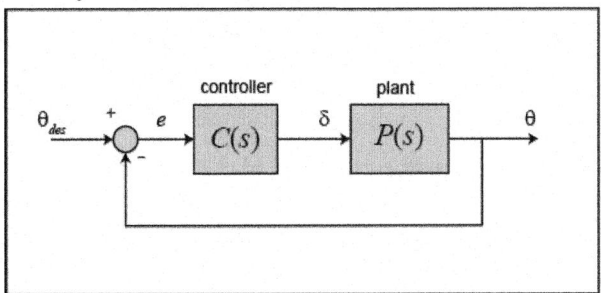

In particular, we will take advantage of the automated tuning capabilities of the SISO Design Tool within MATLAB to design our PID controller. First, enter the following code at the command line to define the model of our plant $P(s)$. Refer to the Aircraft Pitch: System Modeling page for the details of getting these commands.

```
s = tf('s');
P_pitch = (1.151*s+0.1774)/(s^3+0.739*s^2+0.921*s);
```

Proportional control

Let's begin by designing a proportional controller of the form $C(s) = Kp$. The SISO Design Tool we will use for design can be opened by typing `sisotool(P_pitch)` at the command line. This will open both the SISO Design Task window as well as the Control and Estimation Tools Manager window. The SISO Design Task window will initially open with a root locus and Bode plot for the provided plant transfer function and can be employed for graphically tuning a controller. Since we are going to apply the automated tuning function of the SISO Design Tool, you may close the SISO Design Task window.

The Control and Estimation Tools Manager window displays the architecture of the control system being designed as shown below. This default agrees with the architecture we are employing.

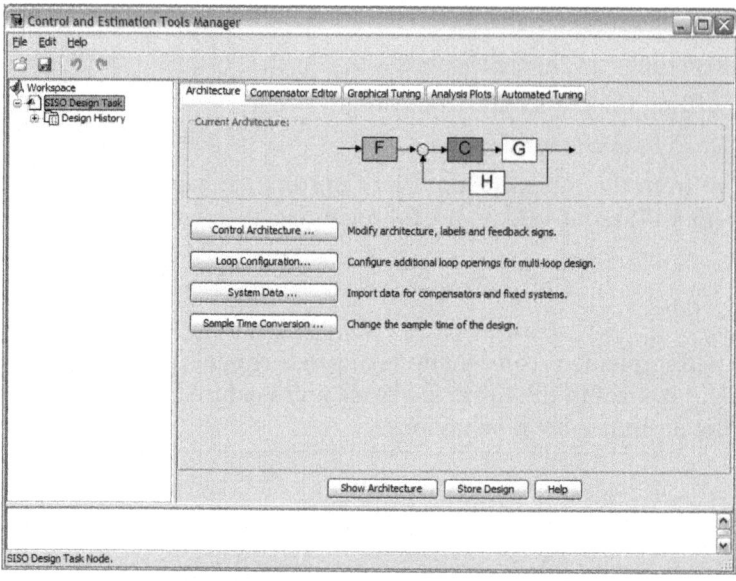

Since our reference is a step function of 0.2 radians, we can set the precompensator block $F(s)$ equal to 0.2 to scale a unit step input to our system. This can be accomplished from the Compensator Editor tab of the open window. Specifically, choose F from the drop-down menu in the Compensator portion of the window and set the compensator equal to 0.2 as shown in the figure below.

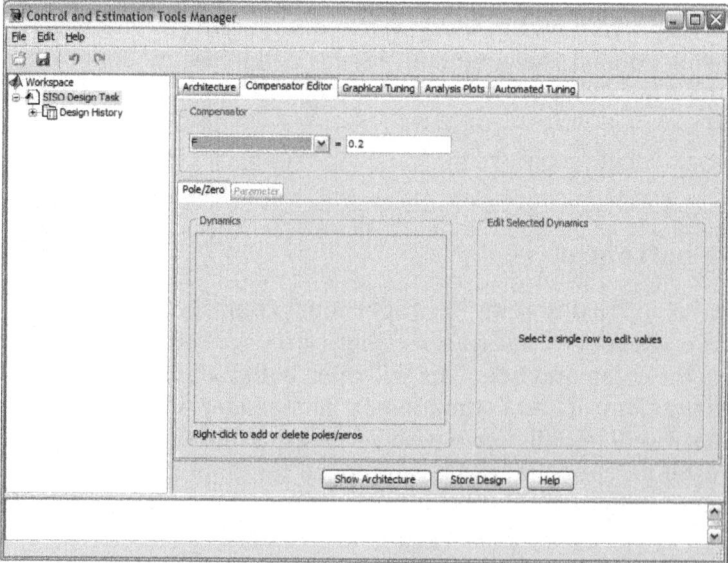

To begin with, let's see how the system performs with a proportional controller Kp set equal to 2. The compensator $C(s)$ can be defined in the same manner as the precompensator, just choose C from from the drop-down menu in the Compensator portion of the window instead of F. Then set the compensator equal to 2. To see the performance of our system with this controller, go to the Analysis Plots tab of the Control and Estimation Tools Manager window. Then choose a Plot Type of Step for Plot 1 in the Analysis Plots section of the window as shown below. Then choose a response of Closed loop r to y for Plot 1 as shown in the figure below.

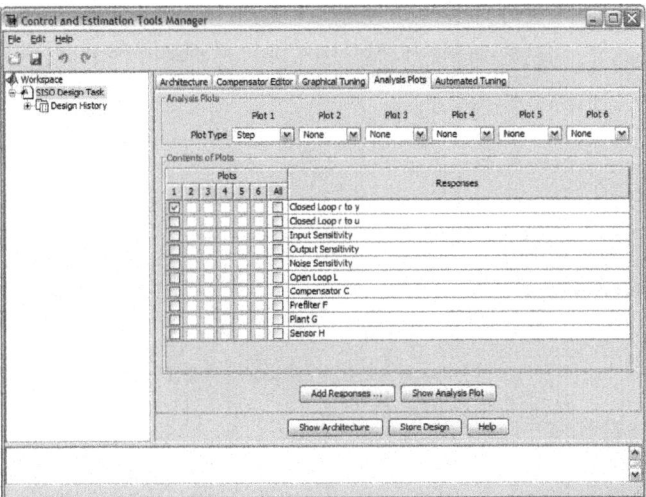

A window will then open with the following step response displayed.

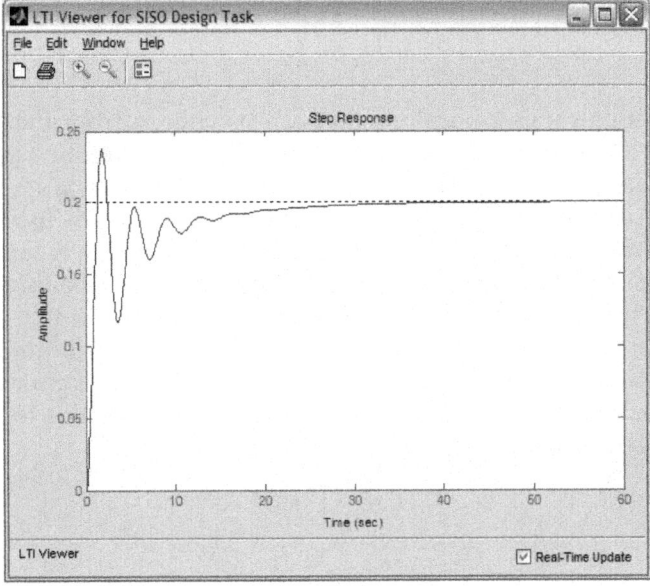

Examination of the above shows that aside from steady-state error, the given design requirements have not been met. The gain chosen for K_p can be adjusted in an attempt to modify the resulting performance through the Compensator Editor tab. Instead, we will use the SISO Design Tool to automatically tune our proportional compensator. In order to use this feature, go to the Automated Tuning tab and choose PID Tuning from the Design method drop-down menu. Then select a Controller type of P as shown in the figure below.

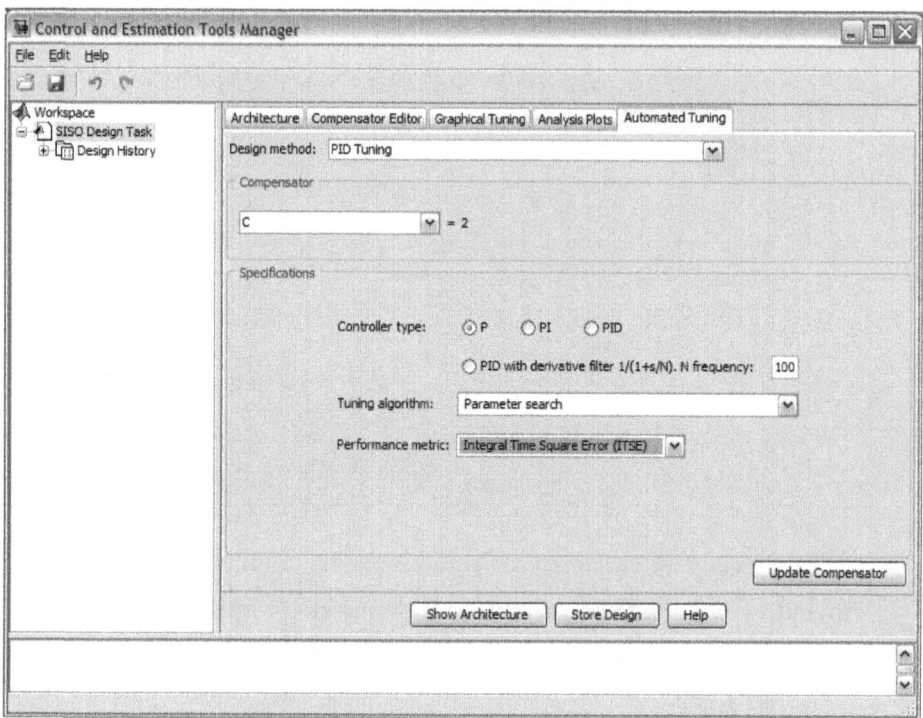

There are a range of options that can be chosen from the Tuning algorithm drop-down menu. These options range from heuristic techniques, like Ziegler-Nichols, to numerical approaches that search over all possible control gains to minimize some identified performance index that is usually related to the error $e(t)$ in the control system. As these metrics attempt to minimize the error over time, they address both transient requirements such as overshoot and rise time, as well as steady-state requirements such as steady-state error. For our example, choose Parameter search from the Tuning algorithm menu. Then choose Integral Time Square Error (ITSE) as the Performance metric. This approach will search over a range of proportional gains to find one that minimizes the following metric.

$$ITSE = \int_0^T te(t)^2\, dt$$

In order to be useful, a metric must always be greater than zero. This can be achieved by taking the absolute value of the error, or by squaring the error as is done in this case. An added effect of taking the square of the error is that large errors are penalized more heavily than small errors. The ITSE metric also multiplies the squared error by the time t. The effect of this is to minimize the contribution of a large initial error due to the step reference.

Once all of the tuning settings have been chosen, you then click the Update Compensator button. This algorithm found that a proportional gain of $K_p = 0.01854$ minimizes the ITSE metric leading to the following closed-loop step response.

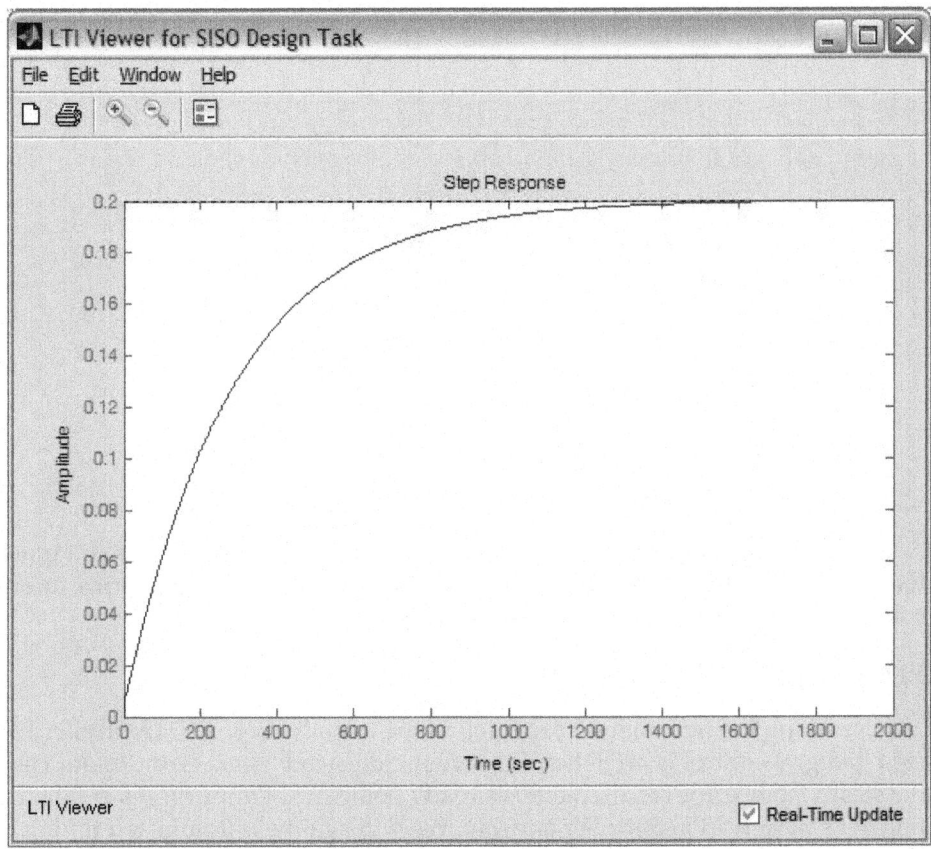

The resulting response is very slow, even slower than our somewhat randomly chosen initial gain $Kp = 2$. This seems to be related to the fact that our metric weighs error at larger values of t more heavily. Therefore, try changing the Performance metric to Integral Square Error (ISE) and click the Update Compensator button again. The gain that minimizes this metric is found to be $Kp = 1$ and the resulting closed-loop step response is shown below.

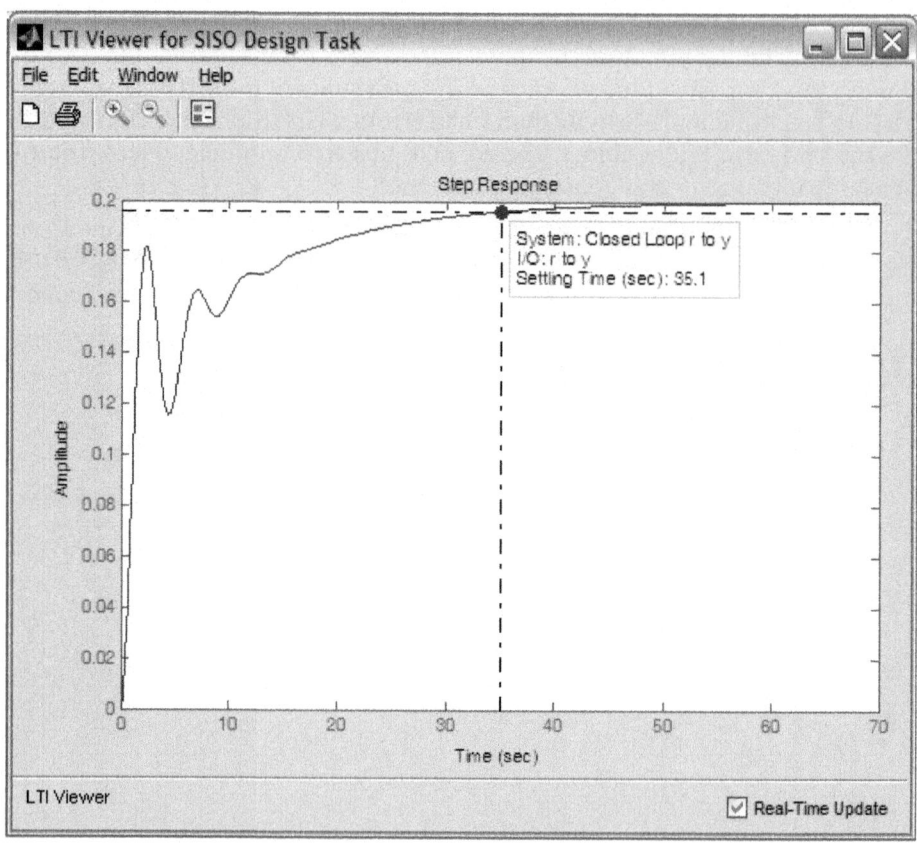

The resulting performance is improved, though the settle time is still much too large. We will likely need to add integral and/or derivative terms to our controller in order to meet the given requirements.

PI control

Recalling the information provided in the Introduction: PID Controller Design integral control is often helpful in reducing steady-state error. In our case, the steady-state error requirement is already being met. For purposes of illustration, let's design a PI controller anyway. We will again use automated tuning to choose our controller gains. Under the Automated Tuning tab change the Controller type to PI and leave the Tuning algorithm as Parameter search and the Performance metric as Integral Square Error (ISE). Clicking on the Compensator Update button then produces the following controller.

$$C(s) = 0.55599 \frac{1+1.8s}{s} \approx \frac{0.56}{s} + 1.00$$

This transfer function is a PI compensator with $K_i = 0.56$ and $K_p = 1.00$. The resulting closed-loop step response is shown below.

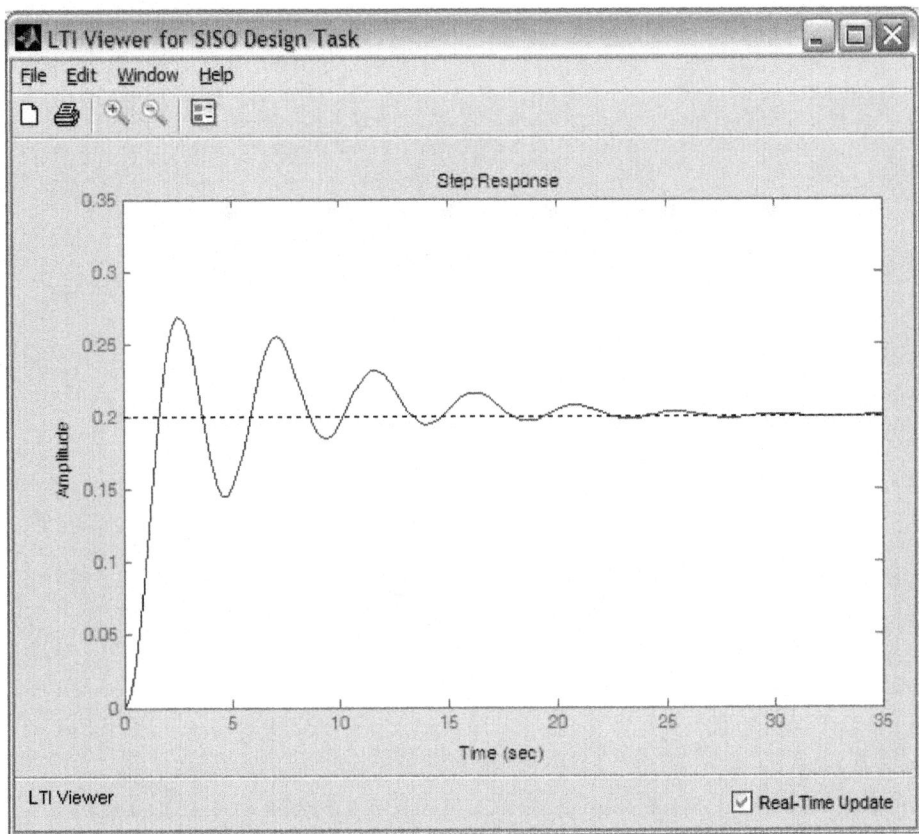

From inspection of the above, notice that the addition of integral control helped reduce the average error in the signal more quickly. Unfortunately, the integral control also made the response more oscillatory, therefore, the settle time requirement is still not met. Furthermore, the overshoot requirement is no longer met either. Let's try also adding a derivative term to our controller.

PID Control

Again recalling the lessons we have learned in the Introduction: PID Controller Design increasing the derivative gain Kd in a PID controller can often help reduce overshoot. Therefore, by adding derivative control we may be able to reduce the oscillation in the response a sufficient amount that we can then increase the other gains to reduce the settling time. Let's test our hypothesis by changing the Controller type to PID and again clicking the Update Compensator button. The generated controller is shown below.

$$C(s) = 4.4545 \frac{1 + 0.22s + 1.1s^2}{s} \approx \frac{4.45}{s} + 0.98 + 4.90s$$

This transfer function is a PID compensator with $K_i = 4.45$, $K_p = 0.98$, and $K_d = 4.90$. The resulting closed-loop step response is shown below.

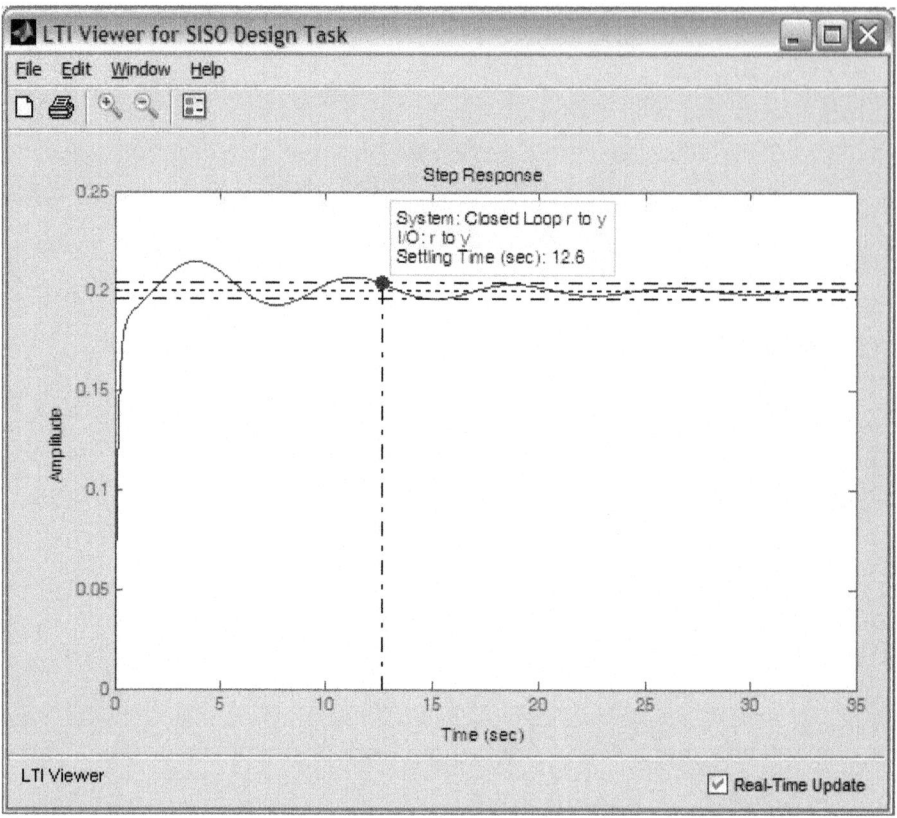

This response meets all of the requirements except for the settle time which at 12.6 seconds is a little larger than the given requirement of 10 seconds. We will attempt to increase the proportional gain in order to reduce the system's settle time. Increasing K_p means that we will no longer achieve the minimum possible performance metric, however, we are willing to do that in order to decrease the resulting settle time. Specifically, we will change K_p so that it equals 2. The resulting PID controller is shown below.

$$C(s) = 4.4545 \frac{1 + 0.45s + 1.1s^2}{s} \approx \frac{4.45}{s} + 2.00 + 4.90s$$

In order to see the effect of this compensator on the closed-loop step response, you need to modify the compensator $C(s)$ manually. This can be done under the Compensator Editor tab of the Control and Estimation Tools Manager window. Specifically, the Complex Zero of the compensator needs to be modified so that it has a Real part of -0.2041 and an Imaginary part of 0.9314 corresponding to the numerator of our controller shown above. Once these changes are made, the step response should appear as follows.

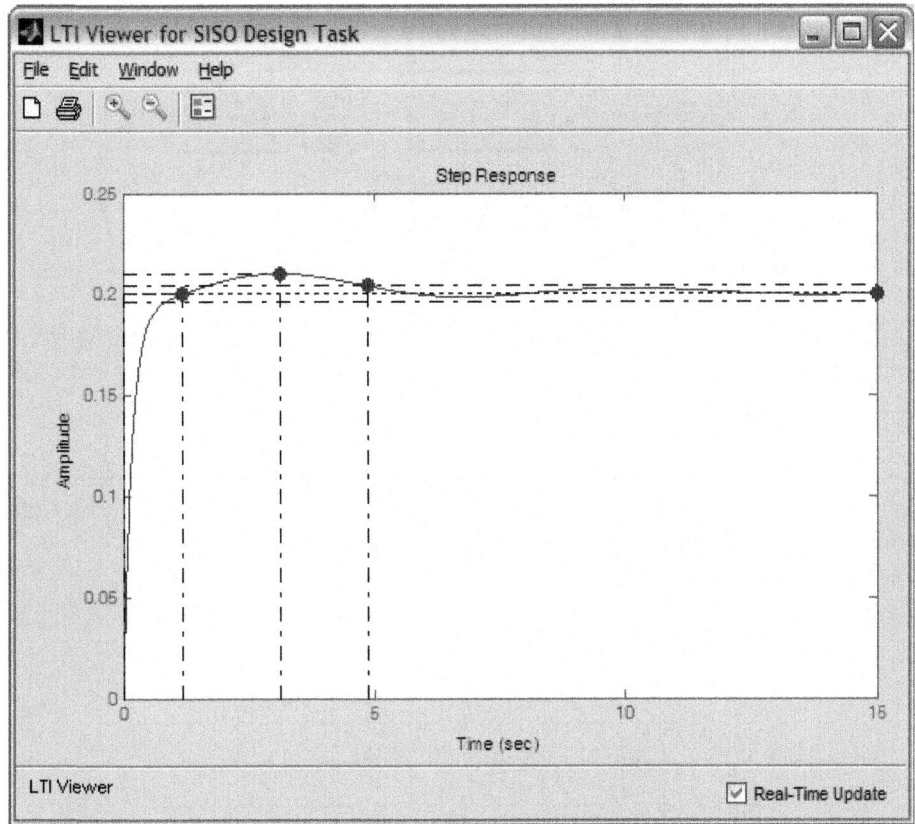

The response shown above meets all of the given requirements as summarized below.

- Overshoot = 5% < 10%
- Rise time = 1.2 seconds < 2 seconds
- Settling time = 5 seconds < 10 seconds
- Steady-state error = 0% < 2%

Therefore, this PID controller will provide the desired performance of the aircraft's pitch.

QUADCOPTER PID EXPLAINED AND TUNING

Many Multicopter, Quadcopter software (such as KK2.0, Multiwii, etc) allow users to change PID configuration values to adjust the performance of their quadcopters. In this post i will try to explain briefly what is PID, how does it affect the stability of the quadcopter (or multicopter), and how to tune the PID for your quadcopter.

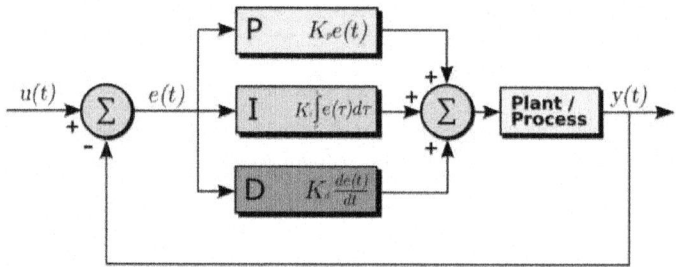

Apart from PID, Rates and expo also affects your flight performance and control.

What is PID?

PID (proportional-integral-derivative) is a closed-loop control system that try to get the actual result closer to the desired result by adjusting the input. Quad-copters or multicopters use PID controller to achieve stability.

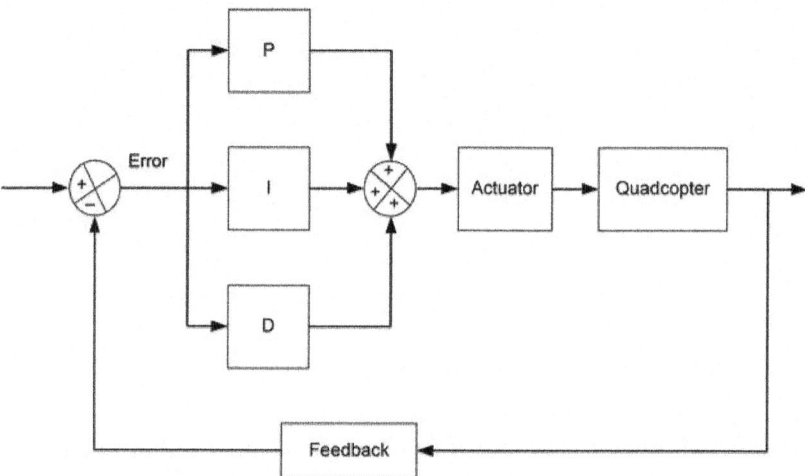

There are 3 algorithms in a PID controller, they are P, I, and D respectively. P depends on the present error, I on the accumulation of past errors, and D is a prediction of future errors, based on current rate of change. These controller algorithms are translated into software code lines.

To have any kind of control over the quadcopter or multicopter, we need to be able to measure the quadcopter sensor output (for example the pitch angle), so we can estimate the error (how far we are from the the desired pitch angle, *e.g.* horizontal, 0 degree). We can then apply the 3 control algorithms to the error, to get the next outputs for the motors aiming to correct the error.

You don't need to understand how PID controller works in order to fly a quadcopter. However, if you want more theory, here is a very interesting and detail explanation of PID controller with examples. This PID is also very good and easy to understand for beginners.

There are three parameters that a pilot can adjust to improve better quadcopter stability. These are the coefficients to the 3 algorithms we mentioned above. The coefficent basically would change the importance and influence of each algorithm to the output. Here we are going to look at what are the effects of these parameters to the stability of a quadcopter.

Per Axis PID structure

Rotational rate
from gyro

actual angle
from sensors

Stabilise
PID

desired
rotational rate

Rate
PID

→ Motor output

desired
angle from
pilot

The Effect of Each Parameter

The variation of each of these parameters alters the effectiveness of the stabilization. Generally there are 3 PID loops with their own P I D coefficients, one per axis, so you will have to set P, I and D values for each axis (pitch, roll and yaw).

To a quadcopter, these parameters can cause these behaviour.

- **Proportional Gain coefficient** – you quadcopter can fly relatively stable without other parameters but this one. This coefficient determines which is more important, human control or the values measured by the gyroscopes. The higher the coefficient, the higher the quadcopter seems more sensitive and reactive to angular change. If it is too low, the quadcopter will appear sluggish and will be harder to keep steady. You might find the multicopter starts to oscillate with a high frequency when P gain is too high.

- **Integral Gain coefficient** – this coefficient can increase the precision of the angular position. For example when the quadcopter is disturbed and its angle changes 20 degrees, in theory it remembers how much the angle has changed and will return 20 degrees. In practice if you make your quadcopter go forward and the force it to stop, the quadcopter will continue for some time to counteract the action. Without this term, the opposition does not last as long. This term is especially useful with irregular wind, and ground effect (turbulence from motors). However, when the I value gets too high your quadcopter might begin to have slow reaction and a decrease effect of the Proportional gain as consequence, it will also start to oscillate like having high P gain, but with a lower frequency.

- **Derivative Gain coefficient** – this coefficient allows the quadcopter to reach more quickly the desired attitude. Some people call it the accelerator parameter because it amplifies the user input. It also decrease control action fast when the error is decreasing fast. In practice it will increase the reaction speed and in certain cases an increase the effect of the P gain.

THE BASICS OF TUNING PID LOOPS

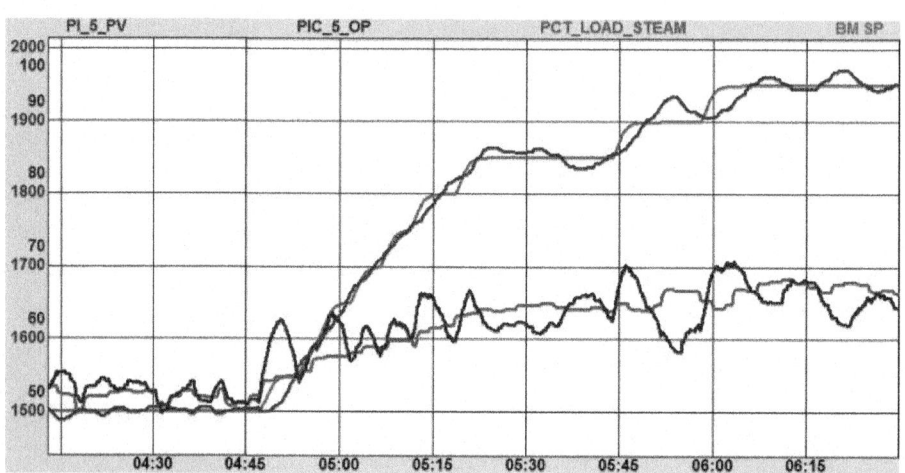

The art of **tuning a PID loop** is to have it adjust its OP to move the PV as quickly as possible to the SP (**responsive**), minimize **overshoot**, and then hold the PV steady at the SP without excessive OP changes (**stable**).

First, Some Definitions

- **PID** = Proportional, Integral, Derivative algorithm. This is not a P&ID, which is a Piping (or Process) and Instrumentation Diagram.
- **PV** = Process Variable - a quantity used as a feedback, typically measured by an instrument. Also sometimes called "MV" - Measured Value.
- **SP** = SetPoint - the desired value for the PV.
- **OP** = OutPut - a signal to a device that can change the PV - frequently a valve, damper, or a pump speed reference. Also sometimes called "CV" - Controlled Value.
- **Overshoot** = when the PV moves further past the SP than desired.
- A PID loop in **manual** (as opposed to **automatic**) only changes its OP upon operator request.
- A loop in **remote** has its SP automatically adjusted by external logic. In **local** the SP is only changed by the operator. Some systems combine auto and remote into "cascade" mode.

- A **direct acting** PID loop increases its OP in response to increasing PV, while a **reverse acting** loop decreases its OP. "Normal" loops are reverse acting. Loops controlling level or pressure via a valve on an output, or temperature via cooling are generally direct acting – "backwards" loops.
- **Error** = the difference between PV and SP.

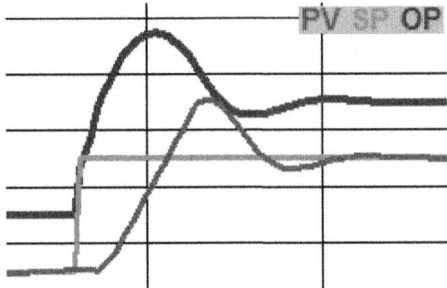

Fig. : Overshoot.

3 Basic Tuning Parameters of a PID Loop

Note: for demonstration purposes the charts below show the individual responses of the actions where the PV is **NOT** affected by the OP. Normally the PV would be affected by the change in OP & would therefore be brought back toward the SP as a result of the OP's response.

Gain: Also called proportional band or P-gain, the gain determines how much change the OP will make due to a change in error (from a PV change and / or an SP change). This mainly corrects the OP based on upsets as they happen. "Gain" implies that a larger number will have more effect. "Proportional band" implies the opposite. P-Gain = 100% / P-band.

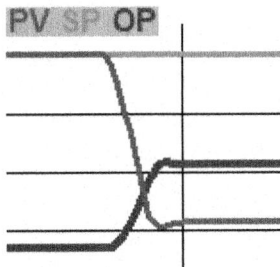

Fig. : Gain Only Response.

Reset: Also called integral or I-gain, the reset determines how much to change the OP over time due to the error (regardless of the direction of movement of the error). This brings a stable PV that is off SP toward the SP. Reset or I-gain implies that a larger number will have more effect. Integral implies the opposite. Reset [resets per minute] = 60 / Integral [seconds per reset].

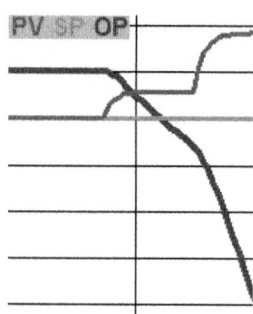

Fig. : Reset Only Response.

Preact

Also called **derivative or D-gain**, the **preact** determines how much to change the OP due from a change in direction of the error or PV. While acting on the PV, rather than the error, is an option in some loops, acting on PV is better because it is undesirable to bump the OP when the SP is changed. It is called preact because it allows the loop to "anticipate" upsets as they begin to happen and react quickly.

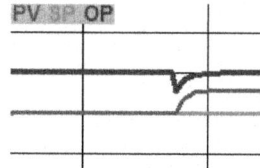

Fig. : Preact Only Response.

Actions Working Together

When the PV is approaching the SP, Proportional and Integral work in opposite directions to cause the PV of a properly tuned loop to get to the SP quickly without excessively overshooting. For a typical reverse-acting loop, the proportional will try to close the OP as the PV rises toward the SP, but the integral will try to open the OP because PV is below SP. As the PV gets closer to the SP, integral action decreases resulting in the PV smoothly decelerating into the SP.

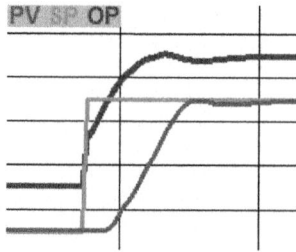

Fig. : Gain and Reset Together.

Initial Loop Tuning

The goal of tuning a PID loop is to make it stable, responsive and to mini-mize overshooting. These goals - especially the last two - conflict with each other. You must find a compromise between the goals which acceptably satisfies them all. Process requirements and physical limitations will determine the balance between amount of acceptable overshoot as well as the demand for responsive-ness. The primary factors which dictate the limits to responsiveness of a loop are:

1. The amount of PV change resulting from an OP step change.

2. The time from when the OP changes until the change starts to be seen in the PV,

3. The time for the PV to reach its new level. Because the PV usually ap-proaches its new level asymptotically, for that time constant we frequently use the time it takes the system's step response to reach 1-1/e, or about 63% of the distance from the initial to its final value.

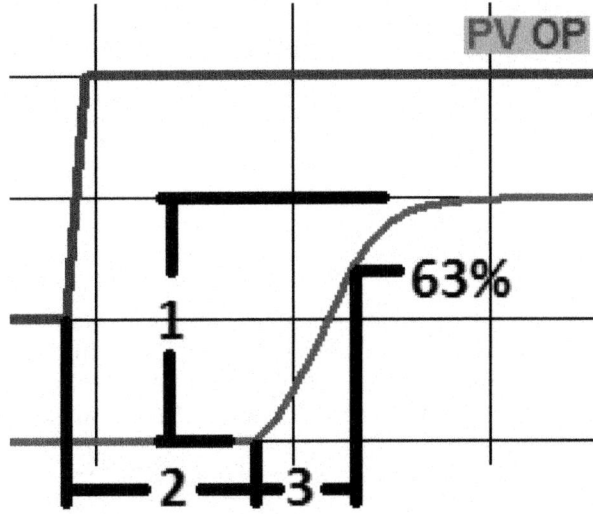

Fig. : Three Response Factors.

When beginning to tune a loop, first make sure you have a good trending package. Watch the PV, SP, OP and other outside variables you suspect might influence the PV together. Characterize the loop by making step changes in manual through the range of the OP and back, and make a note of these three factors. Stepping through the range in both directions is valuable to quantify the **linear-ity** and **hysteresis** of the system.

Linearity: If the same change in OP through the whole scale results in a similar change in PV at each point, the system is **linear**. But if a change on one part of the OP range results in more PV change than the same OP change in a different range, the system is **non-linear**.

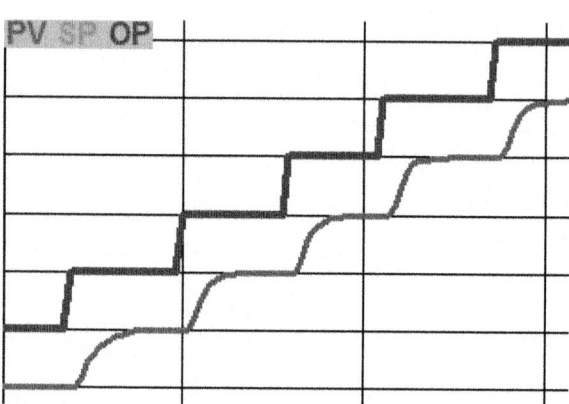

Fig. : Linear OP.

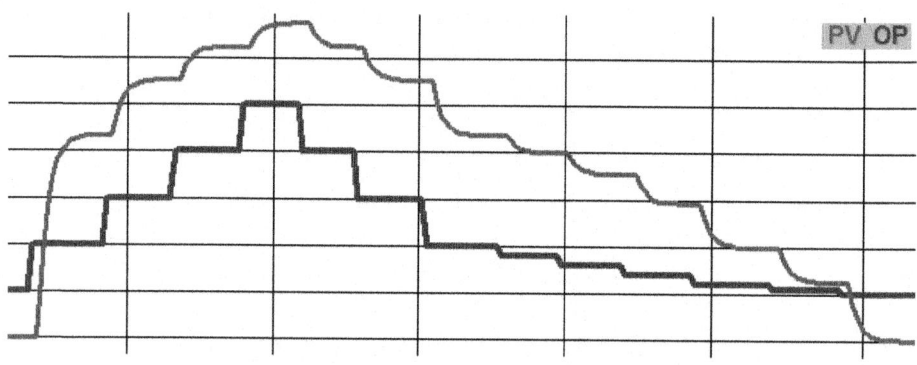

Fig. : Non Linear OP.

Hysteresis: Some devices will yield a different PV for the same OP depending on whether the OP went up or down to get there. A valve might allow 25 GPM through after moving from 20% to 30%, but 30 GPM after moving from 40% to 30%.

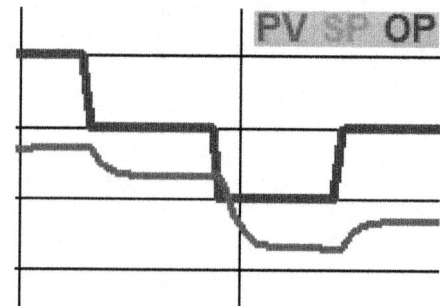

Fig. : OP Hysteresis.

Basic PV Categories: Particle and Bulk

Particle properties are those where a fluid in a pipe may have different properties in different areas, so that the fluid must be mixed or moved to change the property at the PV measurement point. Particle properties include temperature, pH, conductivity, *etc.* These tend to have a significant delay between a change in OP and the beginning of the change in PV (#2 above), and therefore might benefit from derivative action.

Bulk properties describe the state of the fluid as a whole so that it all changes everywhere in a pipe or vessel (for practical purposes) simultaneously. Examples: flow, level, and pressure. Generally these PVs begin to show the result of an OP change immediately (even if the time constant to complete the change is long) and do not need any derivative.

Another categorization of PVs: some (such as flow) increase when the OP increases and decrease when the OP decreases. These should be characterized as shown above - they typically need more integral and minimal P-gain. For others (such as level) the **direction** (rather than value) of the PV is relative to the OP. For the latter, a characterization is more subtle - you want to characterize the slope of the PV for various OPs instead of its value. These sometimes need moderate-to-high gain and less integral.

Starting Parameters

Loops where the PV changes quickly due to a change in OP (flow, or pressure or level in vessels with fast turnover) should have low P-gain (perhaps 0.2) and higher reset (1.5 – 10 rpm). Loops where the PV changes slowly, or changes its direction of movement due to change in OP (temperature and level in vessels with slow turnover) typically need high gain (3 – 100) and low reset (0.05 – 0.3).

These recommended starting parameters are based on the input and output ranges being the same. Some controllers handle tuning parameters based on percent of span, while others do not make this correction. If the spans are different, corrections would have to be made to the parameters themselves. For example, if the flow through a pipe can be from 0 – 10,000 gpm, and you are adjusting the speed of a VFD from 0-100%, the starting gain and reset would need to be 0.004 and 0.02 instead of 0.4 and 2.0.

There are numeric methods where the natural resonant frequency of a system is determined and parameters set accordingly, but **I've found an iterative, intuitive approach to be more useful**:

- Start with a low proportional and no integral or derivative.
- Double the proportional until it begins to oscillate, then halve it.
- Implement a small integral.
- Double the integral until it starts oscillating, then halve it.

That will get the constants close to where they need to be for fine adjustment. Don't hesitate to put the loop back in manual if the loop goes crazy or while studying the trend.

Fine Tuning

To achieve the goal of a responsive and stable loop with minimal overshoot, the tuning must be tested in response to upsets and at steady state. Upsets can be induced by:

- Changing the SP.
- Putting the loop in manual, changing the OP, then returning to automatic.
- Externally causing a change to the PV.

Once the PV has stabilized at its SP, upset it by stepping the SP. Even a very small change is useful here. The proportional will cause an immediate jump in OP, then the integral will cause the OP to continue ramping in the same direction. When the PV starts to move, the proportional will cause the OP to move back the other way, and the integral action will diminish as the PV approaches the SP. Overshoot is often caused by too much integral and/or not enough proportional.

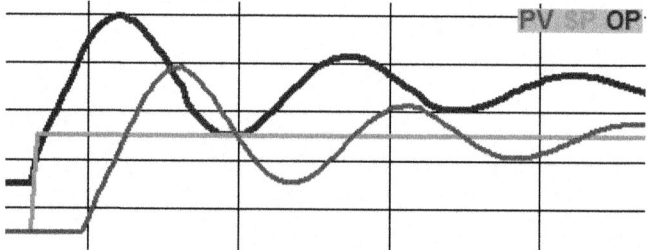

Fig. : Too Much Reset.

The OP needs to start moving back the other way well before the PV reaches the SP. The amount of time between the peak and the PV hitting the SP depends on the nature of the loop. If the peak comes too late, you need more proportional or less integral. If the peak comes too early, you need less proportional or more integral. An early peak will result in the PV leveling out before it reaches the SP, causing the OP and PV to swing on the way to their new steady-state values.

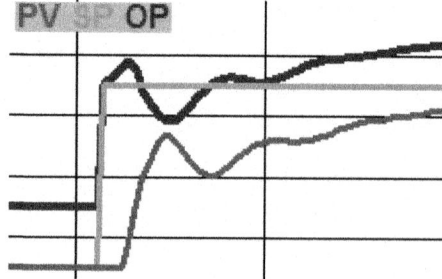

Fig. : Swinging to SP.

Once the loop is roughly tuned, put it in manual and change either the SP or OP, let it stabilize, then put it back in automatic. The loop will then move the PV to the SP, but with integral only - without the initial OP pulse from proportional.

If the OP moves too slowly, you need more integral. A long delay between the change in OP and the end of the resulting change to PV dictates a lower integral value. Introduce derivative if you see that a bump in OP would be beneficial when the PV changes direction at the beginning of an upset.

This is an iterative method – every change in one parameter changes the ideal value for the other parameters. Go back and forth between upset methods and steady state stability, and make sure you check the tuning for the full range of possible SPs, If the system is non-linear, a loop that is stable at higher flows may swing wildly at lower flows, and a loop that is responsive at low flows may be sluggish at higher flows. A PID loop with a control deadband can sometimes achieve acceptable control despite this challenge.

Diagnosing the Cause of Oscillations

If the OP and PV peak at the same time, the oscillation is proportional-driven.

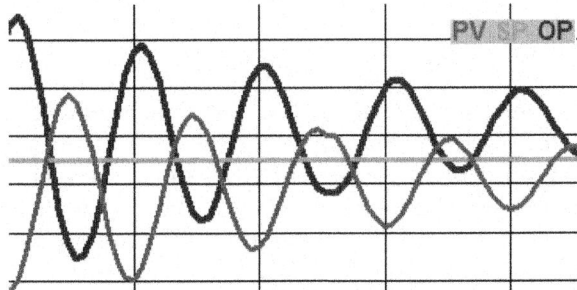

Fig. : Too Much Gain.

If the OP peaks when the PV is crossing its midpoint & visa versa (so that the PV and OP waves are 90° out of phase), the oscillation is integral-driven. Some oscillations are driven by other factors in the system - put the loop in manual to see if it continues to oscillate if you suspect the loop you are tuning is not causing the oscillation.

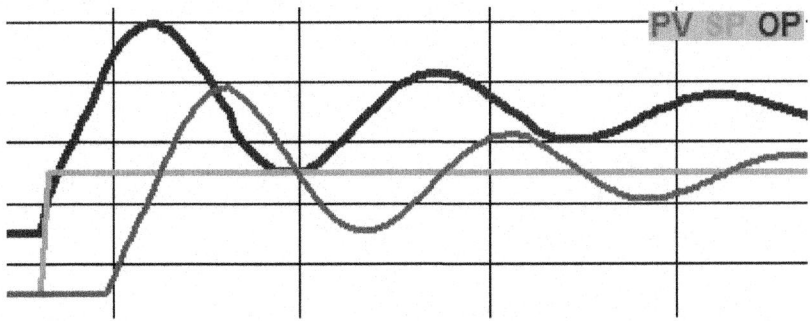

Fig. : Too Much Reset.

Sometimes oscillations are acceptable. For example, the goal of boiler drum level control is primarily to avoid tripping on either low or high level. A moderate amount of oscillation at steady state is a good trade-off to get enough additional responsiveness to avoid tripping following significant upsets.

Note that actuators (especially motorized valves) with deadband and/or limited duty cycle will ALWAYS swing when attached to a traditional PID loop regardless of tuning parameters. You can tell this is happening by looking at the trend – the PV will be flat while the OP is ramping down (due to integral), then the PV jumps to the other side of the SP, and the pattern reverses.

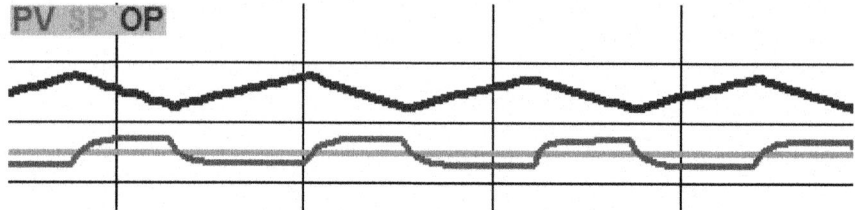

Fig. : Limited Duty Deadband Oscillation.

Balancing the Tuning Goals

A properly tuned loop balances the demands of stability, responsiveness and low overshoot. Tune the loop by adjusting the three tuning parameters so that the loop responds well in a variety of upset and steady-state situations. Some "Advanced PID Tuning Methods" may be necessary if challenges such as non-linearity, deadband, hysteresis or measurable external upsetting events prevent the loop from being satisfactorily tuned with the basic methods described above.

THE P-ONLY CONTROL ALGORITHM

The simplest algorithm in the PID family is a proportional or P-Only controller. Like all automatic controllers, it repeats a measurement-computation-action procedure at every loop sample time, T, following the logic flow shown in the block diagram below:

General Control Loop Block Diagram

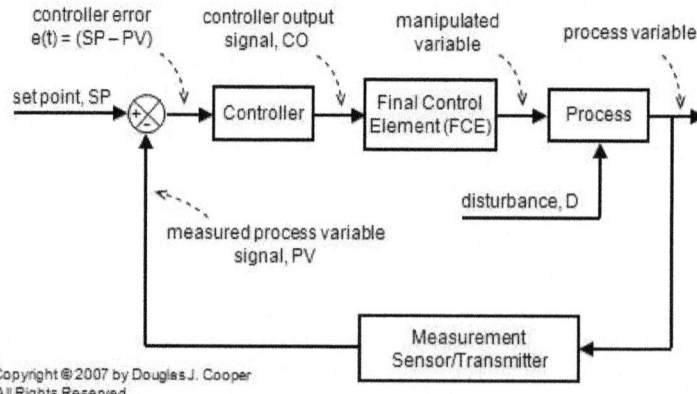

Starting at the far right of the control loop block diagram above:

- A sensor measures and transmits the current value of the process variable, PV, back to the controller (the **'controller wire in'**)

- Controller error at current time t is computed as set point minus measured process variable, or e(t) = SP – PV

- The controller uses this e(t) in a control algorithm to compute a new controller output signal, CO

- The CO signal is sent to the final control element (*e.g.* valve, pump, heater, fan) causing it to change (the **'controller wire out'**)

- The change in the final control element (FCE) causes a change in a manipulated variable

- The change in the manipulated variable (*e.g.* flow rate of liquid or gas) causes a change in the PV

The goal of the controller is to make e(t) = 0 in spite of unplanned and unmeasured disturbances. Since e(t) = SP – PV, this is the same as saying a controller seeks to make PV = SP.

The P-Only Algorithm

The P-Only controller computes a CO action every loop sample time T as:

$$CO = CO_{bias} + Kc \cdot e(t)$$

Where:

CO_{bias} = controller bias or null value

Kc = controller gain, *a tuning parameter*

e(t) = controller error = SP – PV

SP = set point

PV = measured process variable

Design Level of Operation

Real processes display a **nonlinear behaviour**, which means their apparent process gain, time constant and/or dead time changes as operating level changes and as major disturbances change. Since controller design and tuning is based on these Kp, Tp and Θp values, controllers should be designed and tuned for a pre-defined level of operation.

When designing a **cruise control system** for a car, for example, would it make sense for us to perform bump tests to generate dynamic data when the car is traveling twice the normal speed limit while going down hill on a windy day? Of course not.

Bump test data should be collected as close as practical to the design PV when the disturbances are quiet and near their typical values. Thus, the design level of operation for a cruise control system is when the car is traveling at highway speed on flat ground on a calm day.

Definition: the design level of operation (DLO) is where we expect the SP and PV will be during normal operation while the important disturbances are quiet and at their expected or typical values.

Understanding Controller Bias

Let's suppose the P-Only control algorithm shown above is used for cruise control in an automobile and CO is the throttle signal adjusting the flow of fuel to the engine.

Let's also suppose that the speed SP is 70 and the measured PV is also 70 (units can be mph or kph depending on where you live in the world). Since PV = SP, then e(t) = 0 and the algorithm reduces to:

$$CO = CO_{bias} + Kc \cdot (0) = CO_{bias}$$

If CO_{bias} is zero, then when set point equals measurement, the above equation says that the throttle signal, CO, is also zero. This makes no sense. Clearly if the car is traveling 70 kph, then some baseline flow of fuel is going to the engine.

This baseline value of the CO is called the bias or null value. In this example, CO_{bias} is the flow of fuel that, in manual mode, causes the car to travel the design speed of 70 kph when on flat ground on a calm day.

Definition: CO_{bias} is the value of the CO that, in manual mode, causes the PV to steady at the DLO while the major disturbances are quiet and at their normal or expected values.

A P-Only controller bias (sometimes called null value) is assigned a value as part of the controller design and remains fixed once the controller is put in automatic.

Controller Gain, Kc

The P-Only controller has the advantage of having only one adjustable or tuning parameter, Kc, that defines how active or aggressive the CO will move in response to changes in controller error, e(t).

For a given value of e(t) in the P-Only algorithm above, if Kc is small, then the amount added to CO_{bias} is small and the controller response will be slow or sluggish. If Kc is large, then the amount added to CO_{bias} is large and the controller response will be fast or aggressive.

Thus, Kc can be adjusted or tuned for each process to make the controller more or less active in its actions when measurement does not equal set point.

P-Only Controller Design

All controllers from the family of PID algorithms (P-Only, PI, PID) should be designed and tuned using our **proven recipe**:

1. Establish the design level of operation (the normal or expected values for set point and major disturbances).

2. Bump the process and collect controller output (CO) to process variable (PV) dynamic process data around this design level.

3. Approximate the process data behaviour with a first order plus dead time (FOPDT) dynamic model.

4. Use the model parameters from step 3 in rules and correlations to complete the controller design and tuning.

The Internal Model Control (IMC) tuning correlations that work so well for **PI** and **PID** controllers cannot be derived for the simple P-Only controller form. The next best choice is to use the widely-published integral of time-weighted absolute error (ITAE) tuning correlation:

Moderate P-Only:

$$Kc = \frac{0.2}{Kp}\left(\frac{Tp}{\theta p}\right)^{1.22}$$

This correlation is useful in that it reliably yields a moderate Kc value. In fact, some practitioners find that the ITAE Kc value provides a response performance so predictably modest that they automatically start with an aggressive P-Only tuning, defined here as two and a half times the ITAE value:

Aggressive P-Only: Kc = 2.5 (Moderate Kc)

Reverse Acting, Direct Acting and Control Action

Time constant, Tp, and dead time, Θp, cannot affect the sign of Kc because they mark the passage of time and must always be positive. The above tuning correlation thus implies that Kc must always have the same sign as the process gain, Kp.

When CO increases on a process that has a positive Kp, the PV will increase in response. The process is direct acting. Given this CO to PV relationship, when in automatic mode (closed loop), if the PV starts drifting too high above set point, the controller must decrease CO to correct the error.

This "opposite to the problem" reaction is called *negative feedback* and forms the basis of stable control.

A process with a positive Kp is direct acting. With negative feedback, the controller must be reverse acting for stable control. Conversely, when Kp is negative (a reverse acting process), the controller must be direct acting for stable control.

Since Kp and Kc always have the same sign for a particular process and stable control requires negative feedback, then:

- direct acting process (Kp and Kc positive) → use a reverse acting controller
- reverse acting process (Kp and Kc negative) → use a direct acting controller

In most commercial controllers, a positive value of the Kc is always entered. The sign (or action) of the controller is then assigned by specifying that the controller is either reverse or direct acting to indicate a positive or negative Kc respectively.

If the wrong control action is entered, the controller will quickly drive the final control element (*e.g.*, valve, pump, compressor) to full on/open or full off/closed and remain there until the proper control action entry is made.

Proportional Band

Some manufacturers use **different forms for the same tuning parameter.** The popular alternative to Kc found in the marketplace is proportional band, PB.

In many industry applications, both the CO and PV are expressed in units of percent. Given that a controller output signal ranges from a minimum (CO_{min}) to maximum (CO_{max}) value, then:

$$PB = (CO_{max} - CO_{min})/Kc$$

When CO and PV have units of percent and both range from 0% to 100%, the much published conversion between controller gain and proportional band results:

$$PB = 100/Kc$$

Many case studies on this site assign engineering units to the measured PV because plant software has made the task of unit conversions straightforward. If this is true in your plant, take care when using these conversion formula.

Implementation Issues

Implementation of a P-Only controller is reasonably straightforward, but this simple algorithm exhibits a phenomenon called "offset." In most industrial applications, offset is considered an unacceptable weakness. We explore P-Only control, offset and other issues for the **heat exchanger** and the **gravity drained tanks** processes.

USING SIGNAL FILTERS IN OUR PID LOOP

In our study of the derivative mode of a PID controller, we explored how noise or random error in the measured process variable (PV) can degrade controller performance.

The derivative action can cause the noise in the PV measurement to be reflected and amplified in the controller output (CO) signal, producing "chatter" in the final control element (FCE). This extreme control action will increase the wear on a mechanical FCE (*e.g.*, a valve) and lead to increased maintenance needs.

Sources of Noise

Random behaviour in the PV measurement arises because of signal noise and process noise.

Signal noise tends to have higher frequency relative to the characteristic dynamics of process control applications (*i.e.*, processes with streams comprised of liquids, gases, powders, slurries and melts). Sources of signal noise include:

- electrical interference
- jitter (clock related irregularities such as variations in sample spacing)
- quantizing of signal samples into overly-broad discrete "buckets" from low resolution or improperly specified instrumentation (*e.g.* too-large measurement span relative to operating range).

Process noise tends to be lower in frequency. This category borders on the philosophical as to what constitutes a disturbance to be controlled versus noise to be filtered.

Bubbles and splashing that randomly corrupts liquid pressure drop measurements is an example of process noise that might benefit from filtering.

A less clear candidate for filtering is a temperature measurement in a poorly-mixed vessel. The mixing patterns can cause lower-frequency random variations in the temperature signal that are unrelated to changes in the bulk vessel temperature.

It is important to emphasize that before we try to filter away a problem, we should first work to understand the source of the random error. Rather than "fix" the noise by hiding it with additional or modified algorithms, we should attempt to reduce or eliminate the problem through normal engineering and maintenance practices.

The Filtered Signal

The plot below shows the random behaviour of a raw (unfiltered) PV signal and the smoother trace of a filtered PV signal.

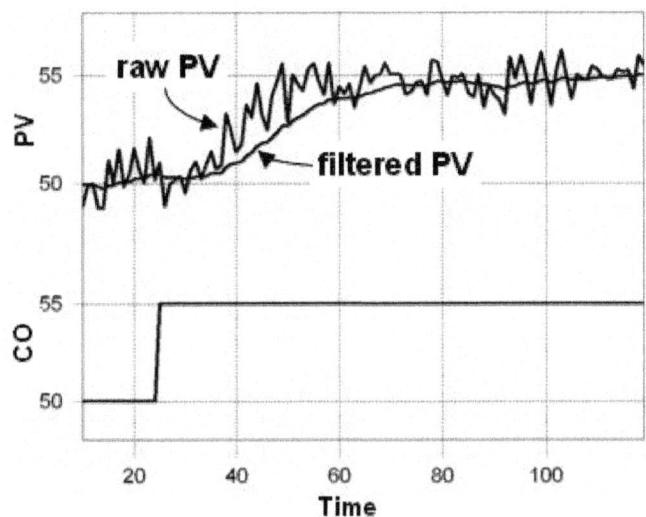

As the above plot illustrates, a filter is able to receive a noisy signal and yield a signal with reduced random variation. A "better" filter design is one that decreases the random variation while retaining more of the true dynamic information of the original signal.

Filters can be analog (hardware) or digital (software); high, low or band pass; linear or nonlinear; designed in the time, Z-transform or frequency domain; and much more. Filters can collect and process data at a rate faster than the control loop sample time, so many data points can go into a single PV sample forwarded to the controller.

Clearly, filter design is a substantial topic. In these articles we offer only an introduction to the basic methods and ideas.

Filters Add Delay

The filtered signal in the plot above, though perhaps visually appealing, clearly lags behind the actual dynamic response of the unfiltered signal. More specifically, the filtered signal has an increased dead time and time constant relative to the behaviour of the actual process.

Signal filters offer benefit in process control applications because they can temper the large CO moves caused by derivative action of noisy PV measurements.

Yet they add delay in sensing the true state of a process, and this has negative consequences in that as delay increases, the best achievable control performance decreases.

The design challenge is to find the careful balance between signal smoothing and information delay to achieve the controller performance we desire for our application.

External Filters in Control

There are three popular places to put external filters in the feedback loop. By "external," we mean that the filters are designed, installed and maintained separately from the controller.

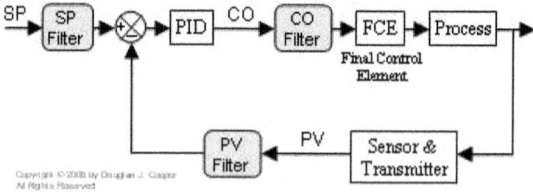

Set Point Filters

Set point filters are not associated with the noisy PV problem.

A set point filter takes a step change in SP, and as shown below, forwards a smooth transition signal to the controller.

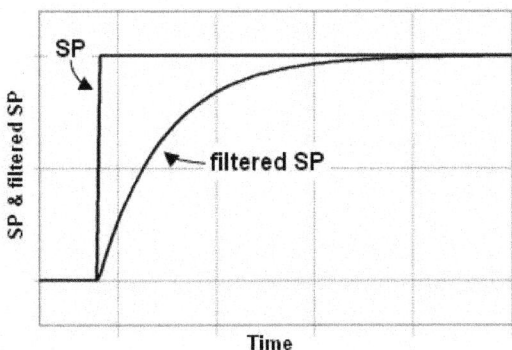

SP filters do not influence the disturbance rejection (regulatory) performance of a controller. Hence, these filters permit a controller to be tuned aggressively to reject disturbances, yet the smoothed SP transition results in a moderate set point tracking (servo) performance from this same aggressive controller.

SP filters are also used to limit overshoot at the top of a set point ramp or step change. If this is our design objective, an alternative is to eliminate the filter and employ a controller that uses proportional on PV rather than proportional on error. For example (compare to PI with proportional on error here):

$$CO = CO_{bias} - KcPV + \frac{Kc}{Ti} \int e(t)dt$$

Finally, and unfortunately, SP filters are occasionally used as a bandage to mask the fact that a controller is simply poorly designed and/or tuned. We all recognize that this is a practice to be avoided.

PV Filters

Signal filters are frequently placed between the sensor transmitter and the controller (or more likely, the multiplexer feeding the controller). In process control applications, these filters should be analog (hardware) devices designed specifically to minimize high frequency electrical interference.

While measurement noise does degrade derivative action in that it leads to chatter in the CO signal, this "noise leads to CO chatter" effect is very modest for proportional action. And interestingly, integral action is unaffected by noise because the constant summing of error literally averages the random variations in the signal.

Since filtering adds delay and this hurts best possible control performance, and because noise is not an issue for proportional and integral action, it is generally poor practice to filter the PV signal external to the controller for anything beyond electrical interference.

The preferred approach is to selectively filter only that signal destined for the derivative computation.

CO Filters

While PV filters smooth the signal feeding the controller, CO filters smooth the noise or "chatter" in the CO signal sent to the final control element. Even if PV signal noise does not appears to cause performance problems, a CO filter can offer potential benefits as it reduces fluctuations in the controller output and this reduces wear on the FCE.

If a noisy PV is an issue in our controller, our first attempts should be to locate and correct the problem. If after that exercise, our decision is to design and implement a filter in our feedback loop, CO filters are attractive alternatives.

Internal Filters in Control

For feedback control, filtering need only be applied to the signal feeding the derivative term. As stated before, noise does not present a problem for proportional and integral action. These elements will perform best without the delay introduced from a signal filter.

When we selectively filter just that signal feeding the derivative calculation, the filter becomes part of the controller architecture. Hence, we can still use our design recipe, though the correlations for tuning this four mode form are different from the four mode "PID with CO Filter" form mentioned above.

There are two common architectures that are identical in capability, though different in presentation.

We filter the PV signal before feeding it to the derivative mode of the PID algorithm for computation:

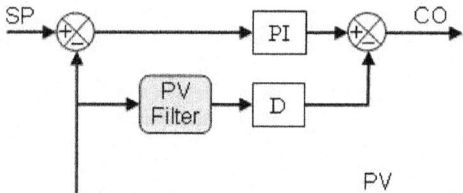

We can also compute the derivative action with the noisy signal and then filter the computed result:

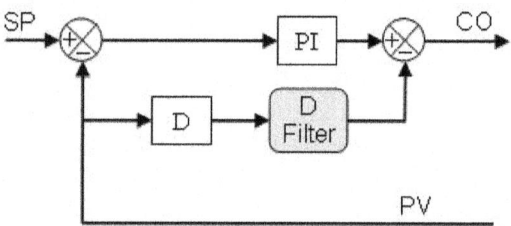

If the same filter form is used (*e.g.*, first-order), we can show mathematically that both options above are identical.

As it turns out, many commercial controllers use this internal derivative filtering form where they implement a first-order filter and fix the filter time at one-tenth of the derivative time value.

GAP CONTROL

Some control loops have two seemingly conflicting objectives of keeping the process variable under control, but also minimizing controller output movement. Although such a loop will have a set point, it is more important to keep its process variable within predefined bounds than to keep it exactly at set point. A typical application of gap control is averaging level control. (Averaging control is somewhat like surge tank control, but the process variable is controlled around its set point. True surge tank level control rarely controls around its set point.)

Controller manufacturers have designed a modification of the standard PID control algorithm for use on processes with these conflicting objectives. This modification is called gap control and it works on the principle of two user-definable control regions, one for each of the two control objective.

The first region is far from set point (outside the gap) and requires a strong control action to turn the process around and bring it back to set point. The normal controller settings are used outside the gap. The second region is close to set point (inside the gap) within which the controller detunes itself based on a configurable gain multiplier M (between 0.0 and 1.0). The detuning helps to minimize controller output movements.

Tuning a Gap Controller

To use gap control effectively, you should set the size of the gap around the set point according to the typical variation of the process variable so that the process variable does not frequently and unnecessarily venture outside the gap. Tune the controller for fast response outside the gap to provide quick recovery from disturbances. Use a gain multiplier (M) of 0.5 or less to minimize the controller output movement inside the gap.

For integrating processes divide the calculated integral time (Ti) by M (*i.e.* make the integral time longer in proportion to decreasing the controller gain). This is a requirement only for integrating processes and is done to ensure a stable integral term when the controller gain is reduced inside the gap. Do not use a gain multiplier of zero when controlling integrating processes because this dead band will cause the process variable to continuously cycle through the gap.

Gap control should not be used in control loops of which the objective is to keep the process variable as close to set point as possible. Use regular PI or PID control for those loops.

CONTROLLING GAPS IS ESSENTIAL TO DEFENSIVE PLAY AND PUCK RECOVERY

The concept of gap control is very important for coaches and players to understand so that they have a common knowledge base when discussing individual situations that occur during games. In a broad sense gap control can be defined as the "distance to the nearest opponent."

The easiest example of gap control occurs on a 1 on 1 situation with the puck carrier approaching a defenseman during an offensive rush. Coaches preach tight gaps to their players and defend the blue line. In practice youth hockey players have a very hard time maintaining a close/tight gap in the neutral ice area and typically will attempt to move to a zero gap when the play reaches a critical point in the scoring area on front of their own net.

A tight gap often is described as one to one and half sticks lengths from the puck carrier. How often do we see it in youth hockey? Not very often because the defenders lack confidence in their skating ability and are afraid they will get beat and then criticized by teammates, coaches and parents.

In reality, playing a tight gap in neutral is more desirable than retreating deep into the defensive zone and getting beat close to the net. If a defender gets beat at the center red line there is plenty of time and space for teammates to fill in up and the mere act of playing a tight gap sooner will slow the opponents attack allowing for more support from the defending team players. Coaches should think about encouraging tight gaps in the neutral zone area and make sure all players know that turnovers are more likely to occur and counter attacks will come quicker.

To be fair to defensemen, they do need to decide if they have a "contain or force" situation to deal with. In a contain scenario, with little or no support or if they are facing an odd man rush, larger gaps are warranted and the main objective is to protect the middle lanes and steer the play to the outside. If support is present a force situation presents itself and the defender can move to zero gap as quickly as possible to create a turnover.

The puck carrier/defender scenario is the most common thought of gap control situation but it is also necessary to expand the concept to scenarios all over the ice. We want zero gap in front of the net most of the time. Supporting forwards should be thinking about tight gaps in their own end of the rink until possession of the puck is regained. Then the other team should be thinking about tight gaps.

A back checking forward should be thinking about his gap with the opposing player and should try to get defensive side positioning with close to a zero gap. Cycling is designed to create gap and defenders can break the cycle by staying with their opponent on a tight gap.

With the speed of today's game and lengthy periods of play without stoppage in play it is essential that as players transition from offense to defense and they maintain good gap control when the other team has possession of the puck.

Gap control applies all over the rink and is an essential part of defending and recovering possession of the puck in order to initiate an offensive play. The concepts with your team and give them permission to run tight gaps and force the play earlier. They will make mistakes but they will quickly learn how effective tight gaps can be.

A good example of all over gap control is evident when watching the Minnesota Wild and other NHL teams. Gap control is an essential element in playing defensively while trying to regain possession of the puck. When someone makes a mistake in this area you will see how it impacts the offense and creates opportunities. Talk to you players about gap control all over the rink.

CONTROLLER TUNING

Controller tuning is the process by which a control engineer or technician selects values of user-adjustable controller parameters (for a PID controller these are the bias, gain, integral time, and derivative time) so that the closed loop dynamic response behaves as desired.

Loop tunings are the primary point of contact between an operations/manufacturing engineer and the plant control system. Controller settings determine the system response: a poorly tuned controller may be as bad as no controller at all.

Tuning is a exercise in compromise. Controller objectives, specifications, requirements, and performance always conflict to some degree or another. There are rarely absolute criteria for selecting tunings and so judgement is required.

As you prepare to tune a loop, you must consider a range of concerns and objectives.

Objectives

All control loops are fundamentally concerned with two objectives: disturbance rejection and setpoint tracking. In the CPI, disturbance rejection is normally the more important concern (despite what the examples in control textbooks may suggest).

An important secondary objective is to minimize the "cost" and variability of your manipulated variables.

Forecasting

Before tuning, you need to have some idea of what to expect. In particular, you want to have an idea of what sort of inputs are likely -- step changes? ramps? impulses? -- and how big they are likely to be. A "tight" tuning designed for small inputs may be the exact opposite of what one would do for a large input.

You also need to understand your system. How much noise do you anticipate? What constraints do safety, the environment, and equipment protection impose on your plans? What constraints do nearby units and equipment impose?

Most methods for obtaining initial tuning sections are based on some sort of model. What type of model are you using? Can you quantify the amount of plant/model mismatch?

Specifications

You can't tune a loop unless you have some way of deciding whether or not it is "working". Consequently, you'll need to determine how you will measure success. Desired response is often quantified in terms of one of the loop performance specifications in the semester. The specifications used depend on the process, but might include:

1. Speed of response
   o Rise time
   o Time to first peak
   o Settling time
2. Oscillation
   o Closed loop damping coefficient (0.4 is a common target)
   o Overshoot
   o Decay ratio (1/4 is common)
   o Frequency or period of oscillation

Loop specifications and performance often interact and conflict. For instance, adding integral action eliminates offset, but tends to slow response time.

General Performance Measures

Sometimes it is useful to use broad measures of performance that focus less on the specifics of the loop than on the general variability and deviation from desired performance. These types of criteria are particularly important in organizations that attempt to measure "quality" and employ statistical quality control techniques. SQC techniques are primarily designed to reduce and eliminate variability.

The *error* in a control loop is usually defined as the deviation from setpoint. There are a variety of ways of quantifying the cumulative error:

- *Integral Error*. The cumulative sum of the error. Specifying an IE value will not ensure a particular type of damping.

$$IE = \int_0^{time} e(t)\, dt$$

- *Integral Absolute Error*. The sum of areas above and below the setpoint, this penalizes all errors equally regardless of direction.

$$IAE = \int_0^{time} |e(t)|\, dt$$

- *Integral Squared Error*. Penalizes large errors more than small.

$$ISE = \int_0^{time} \left(e(t)\right)^2 dt$$

- *Integral Time Weighted Absolute Error*. Penalizes persistant errors.

$$ITAE = \int_0^{time} t\left|e(t)\right| dt$$

- *Integral Time Squared Error*.

$$ITAE = \int_0^{time} t\left|e(t)\right| dt$$

Chapter 6

TUNNING PID CONTROLLERS

INTRODUCTION OF THE PID TUNER

PID Tuner provides a fast and widely applicable single-loop PID tuning method for the Simulink® PID Controller blocks. With this method, you can tune PID controller parameters to achieve a robust design with the desired response time.

A typical design workflow with the PID Tuner involves the following tasks:

(1) Launch the PID Tuner. When launching, the software automatically computes a linear plant model from the Simulink model and designs an initial controller.

(2) Tune the controller in the PID Tuner by manually adjusting design criteria in two design modes. The tuner computes PID parameters that robustly stabilize the system.

(3) Export the parameters of the designed controller back to the PID Controller block and verify controller performance in Simulink.

Opening the Model

Open the engine speed control model with PID Controller block and take a few moments to explore it.

```
open_system('scdspeedctrlpidblock');
```

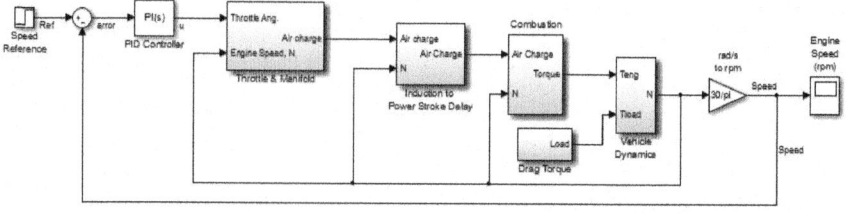

Design Overview

In this example, you design a PI controller in an engine speed control loop. The goal of the design is to track the reference signal from a Simulink step block scd-speedctrlpidblock/Speed Reference. The design requirement are:

• Settling time under 5 seconds

• Zero steady-state error to the step reference input.

In this example, you stabilize the feedback loop and achieve good reference tracking performance by designing the PI controllerscdspeedctrl/PID Controller in the PID Tuner.

Opening the PID Tuner

To launch the PID Tuner, double-click the PID Controller block to open its block dialog. In the Main tab, click Tune.

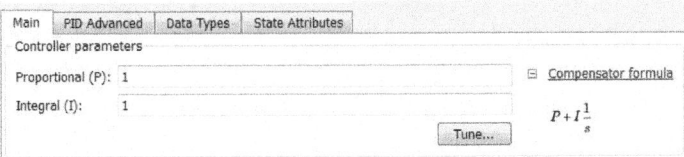

Initial PID Design

When the PID Tuner launches, the software computes a linearized plant model seen by the controller. The software automatically identifies the plant input and output, and uses the current operating point for the linearization. The plant can have any order and can have time delays.

The PID Tuner computes an initial PI controller to achieve a reasonable tradeoff between performance and robustness. By default, step reference tracking performance displays in the plot.

The following figure shows the PID Tuner dialog with the initial design:

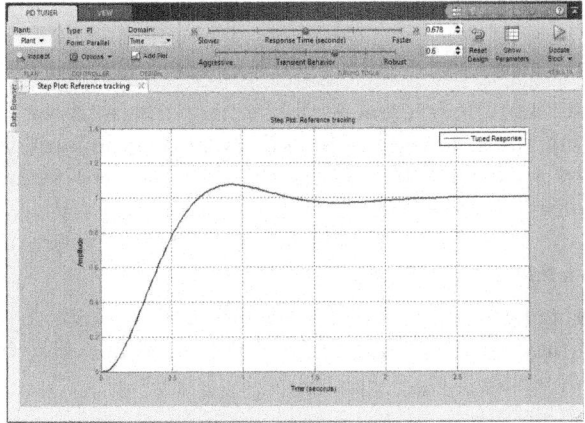

Displaying PID Parameters

Click Show parameters to view controller parameters P and I, and a set of performance and robustness measurements. In this example, the initial PI controller design gives a settling time of 2 seconds, which meets the requirement.

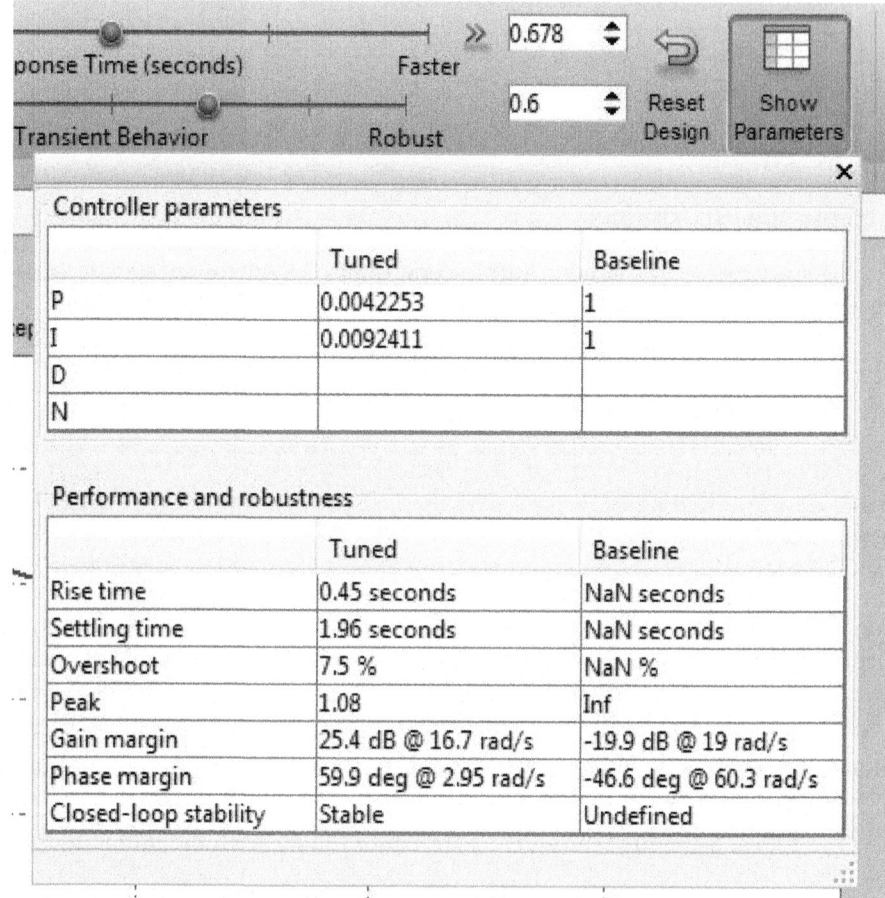

Controller parameters	Tuned	Baseline
P	0.0042253	1
I	0.0092411	1
D		
N		

Performance and robustness	Tuned	Baseline
Rise time	0.45 seconds	NaN seconds
Settling time	1.96 seconds	NaN seconds
Overshoot	7.5 %	NaN %
Peak	1.08	Inf
Gain margin	25.4 dB @ 16.7 rad/s	-19.9 dB @ 19 rad/s
Phase margin	59.9 deg @ 2.95 rad/s	-46.6 deg @ 60.3 rad/s
Closed-loop stability	Stable	Undefined

Adjusting PID Design in the PID Tuner

The overshoot of the reference tracking response is about 7.5 percent. Since we still have some room before reaching the settling time limit, you could reduce the overshoot by increasing the response time. Move the response time slider to the left to increase the closed loop response time. Notice that when you adjust response time, the response plot and the controller parameters and performance measurements update.

The following figure shows an adjusted PID design with an overshoot of zero and a settling time of 4 seconds. The designed controller effectively becomes an integral-only controller.

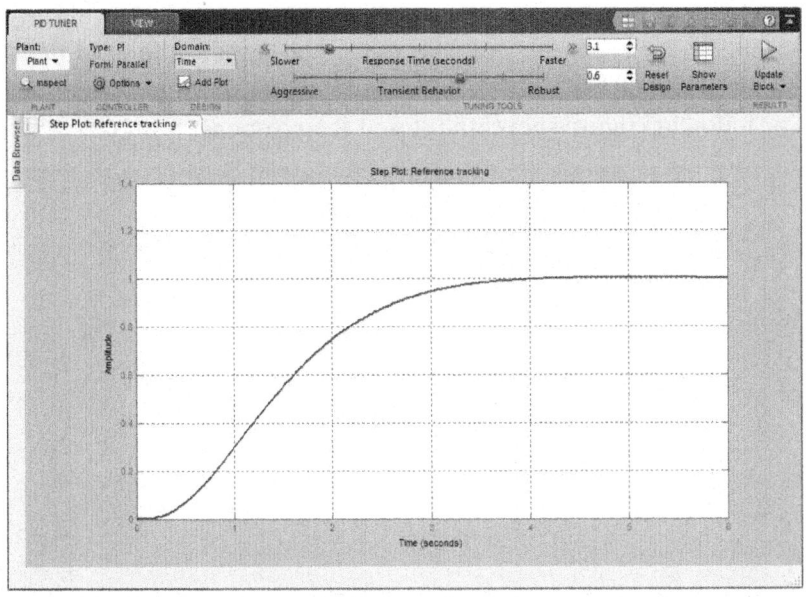

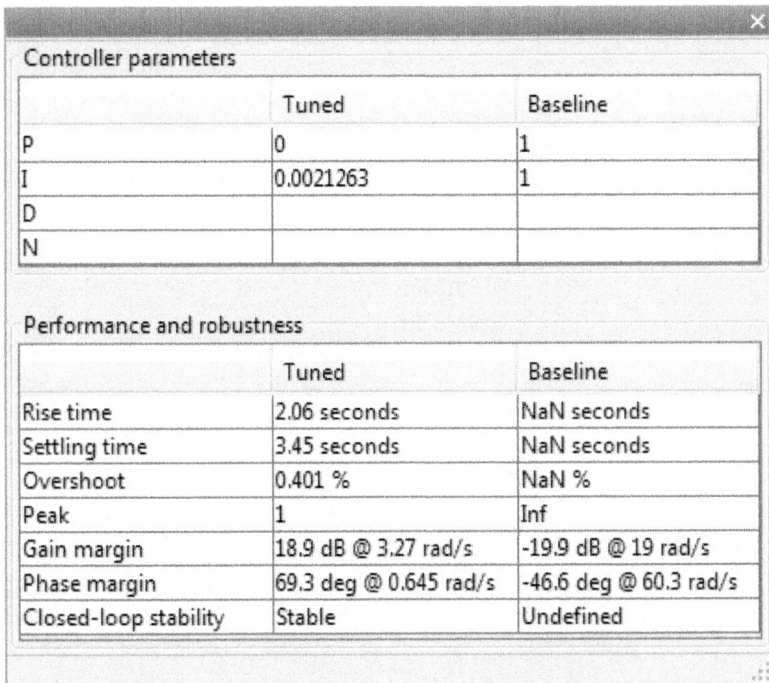

Completing PID Design with Performance Trade-Off

In order to achieve zero overshoot while reducing the settling time below 2 seconds, you need to take advantage of both sliders. You need to make control response faster to reduce the settling time and increase the robustness to reduce the overshoot. For example, you can reduce the response time from 3.4 to 1.5 seconds and increase robustness from 0.6 to 0.72.

The following figure shows the closed-loop response with these settings:

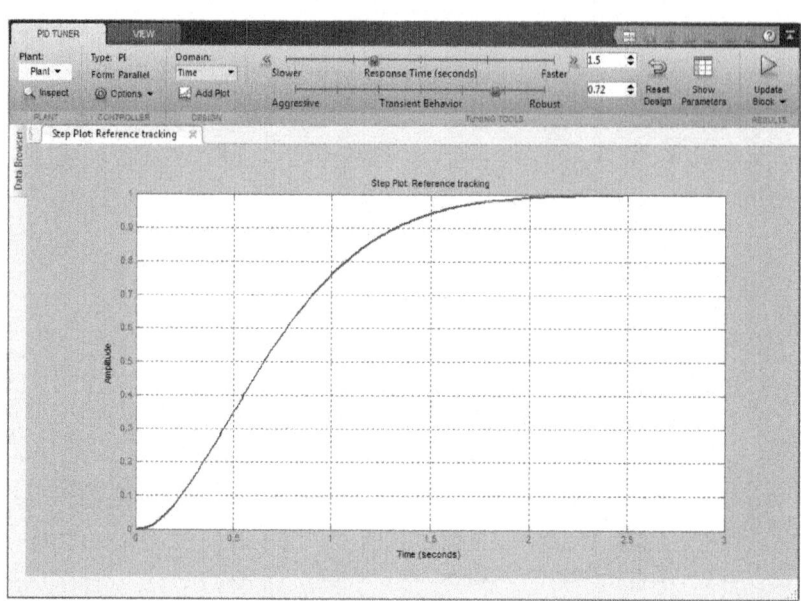

Controller parameters

	Tuned	Baseline
P	0.0014551	1
I	0.0043791	1
D		
N		

Performance and robustness

	Tuned	Baseline
Rise time	1.09 seconds	NaN seconds
Settling time	1.81 seconds	NaN seconds
Overshoot	0 %	NaN %
Peak	0.999	Inf
Gain margin	32.8 dB @ 15 rad/s	-19.9 dB @ 19 rad/s
Phase margin	72 deg @ 1.33 rad/s	-46.6 deg @ 60.3 rad/s
Closed-loop stability	Stable	Undefined

Writing the Tuned Parameters to PID Controller Block

After you are happy with the controller performance on the linear plant model, you can test the design on the nonlinear model. To do this, click Update Block in the PID Tuner. This action writes the parameters back to the PID Controller block in the Simulink model.

The following figure shows the updated PID Controller block dialog:

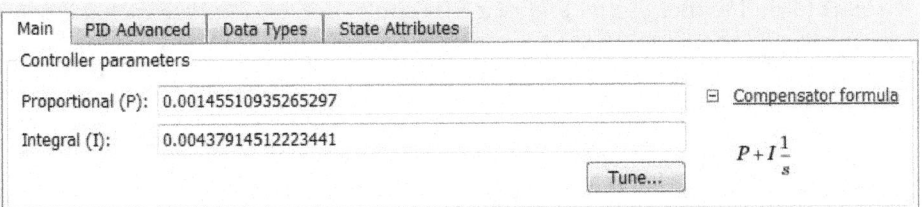

Completed Design

The following figure shows the response of the closed-loop system:

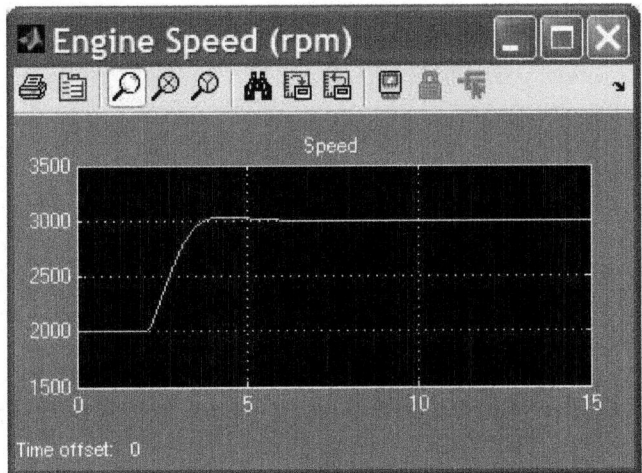

The response shows that the new controller meets all the design requirements.

PID THEORY EXPLAINED

Proportional-Integral-Derivative (PID) control is the most common control algorithm used in industry and has been universally accepted in industrial control. The popularity of PID controllers can be attributed partly to their robust performance in a wide range of operating conditions and partly to their functional simplicity, which allows engineers to operate them in a simple, straightforward manner.

As the name suggests, PID algorithm consists of three basic coefficients; proportional, integral and derivative which are varied to get optimal response.

Closed loop systems, the theory of classical PID and the effects of tuning a closed loop control system. The PID toolset in LabVIEW and the ease of use of these VIs.

Control System

The basic idea behind a PID controller is to read a sensor, then compute the desired actuator output by calculating proportional, integral, and derivative responses and summing those three components to compute the output. Before we start to define the parameters of a PID controller, we shall see what a closed loop system is and some of the terminologies associated with it.

Closed Loop System

In a typical control system, the process variable is the system parameter that needs to be controlled, such as temperature (°C), pressure (psi), or flow rate (liters/minute). A sensor is used to measure the process variable and provide feedback to the control system. The set point is the desired or command value for the process variable, such as 100 degrees Celsius in the case of a temperature control system. At any given moment, the difference between the process variable and the set point is used by the control system algorithm (compensator), to determine the desired actuator output to drive the system (plant). For instance, if the measured temperature process variable is 100 °C and the desired temperature set point is 120 °C, then the actuator output specified by the control algorithm might be to drive a heater. Driving an actuator to turn on a heater causes the system to become warmer, and results in an increase in the temperature process variable. This is called a closed loop control system, because the process of reading sensors to provide constant feedback and calculating the desired actuator output is repeated continuously and at a fixed loop rate.

In many cases, the actuator output is not the only signal that has an effect on the system. For instance, in a temperature chamber there might be a source of cool air that sometimes blows into the chamber and disturbs the temperature.Such a term is referred to asdisturbance. We usually try to design the control system to minimize the effect of disturbances on the process variable.

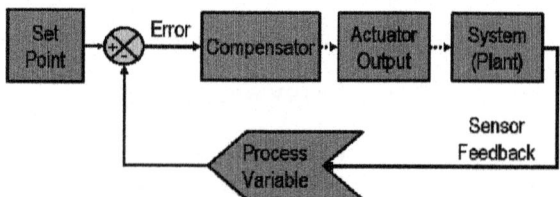

Fig. : Block diagram of a typical closed loop system.

Defintion of Terminlogies

The control design process begins by defining the performance requirements. Control system performance is often measured by applying a step function as the set point command variable, and then measuring the response of the process variable. Commonly, the response is quantified by measuring defined waveform characteristics. Rise Time is the amount of time the system takes to go from 10% to 90% of the steady-state, or final, value. Percent Overshoot is the amount that the process variable overshoots the final value, expressed as a percentage of the final value. Settling time is the time required for the process variable to settle to within a certain percentage (commonly 5%) of the final value. Steady-State Error is the final difference between the process variable and set point. Note that the exact definition of these quantities will vary in industry and academia.

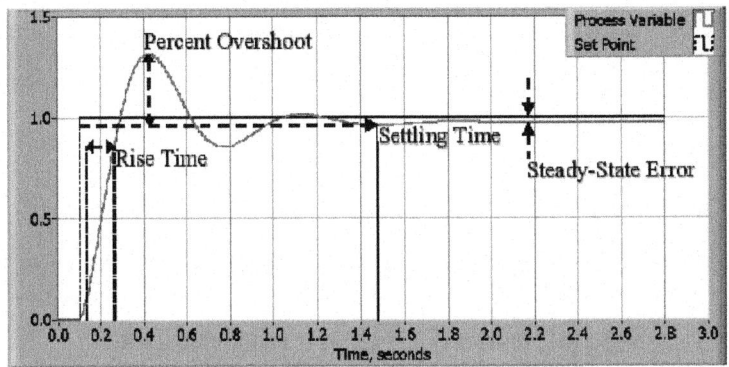

Fig. : Response of a typical PID closed loop system.

After using one or all of these quantities to define the performance requirements for a control system, it is useful to define the worst case conditions in which the control system will be expected to meet these design requirements. Often times, there is a disturbance in the system that affects the process variable or the measurement of the process variable. It is important to design a control system that performs satisfactorily during worst case conditions. The measure of how well the control system is able to overcome the effects of disturbances is referred to as the disturbance rejection of the control system.

In some cases, the response of the system to a given control output may change over time or in relation to some variable. A nonlinear system is a system in which the control parameters that produce a desired response at one operating point might not produce a satisfactory response at another operating point. For instance, a chamber partially filled with fluid will exhibit a much faster response to heater output when nearly empty than it will when nearly full of fluid. The measure of how well the control system will tolerate disturbances and nonlinearities is referred to as the robustness of the control system.

Some systems exhibit an undesirable behaviour called deadtime. Deadtime is a delay between when a process variable changes, and when that change can be observed. For instance, if a temperature sensor is placed far away from a cold water fluid inlet valve, it will not measure a change in temperature immediately if the valve is opened or closed. Deadtime can also be caused by a system or output actuator that is slow to respond to the control command, for instance, a valve that is slow to open or close. A common source of deadtime in chemical plants is the delay caused by the flow of fluid through pipes.

Loop cycle is also an important parameter of a closed loop system. The interval of time between calls to a control algorithm is the loop cycle time. Systems that change quickly or have complex behaviour require faster control loop rates.

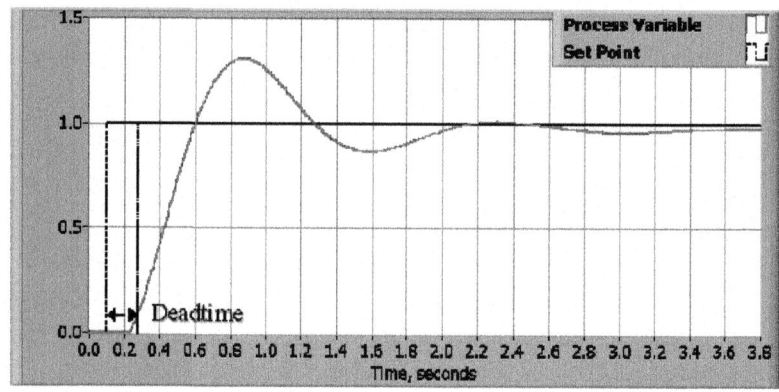

Fig. : Response of a closed loop system with deadtime.

Once the performance requirements have been specified, it is time to examine the system and select an appropriate control scheme. In the vast majority of applications, a PID control will provide the required results

PID Theory

Proportional Response

The proportional component depends only on the difference between the set point and the process variable. This difference is referred to as the Error term. The proportional gain (K_c) determines the ratio of output response to the error signal. For instance, if the error term has a magnitude of 10, a proportional gain of 5 would produce a proportional response of 50. In general, increasing the proportional gain will increase the speed of the control system response. However, if the proportional gain is too large, the process variable will begin to oscillate. If K_c is increased further, the oscillations will become larger and the system will become unstable and may even oscillate out of control.

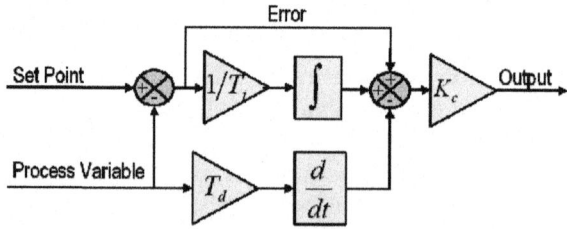

Fig. : Block diagram of a basic PID control algorithm.

Integral Response

The integral component sums the error term over time. The result is that even a small error term will cause the integral component to increase slowly. The integral response will continually increase over time unless the error is zero, so the effect is to drive the Steady-State error to zero. Steady-State error is the final difference between the process variable and set point. A phenomenon called integral windup results when integral action saturates a controller without the controller driving the error signal toward zero.

Derivative Response

The derivative component causes the output to decrease if the process variable is increasing rapidly. The derivative response is proportional to the rate of change of the process variable. Increasing the derivative time (T_d) parameter will cause the control system to react more strongly to changes in the error term and will increase the speed of the overall control system response. Most practical control systems use very small derivative time (T_d), because the Derivative Response is highly sensitive to noise in the process variable signal. If the sensor feedback signal is noisy or if the control loop rate is too slow, the derivative response can make the control system unstable

Tuning

The process of setting the optimal gains for P, I and D to get an ideal response from a control system is called tuning. There are different methods of tuning of which the "guess and check" method and the Ziegler Nichols method.

The gains of a PID controller can be obtained by trial and error method. Once an engineer understands the significance of each gain parameter, this method becomes relatively easy. In this method, the I and D terms are set to zero first and the proportional gain is increased until the output of the loop oscillates. As one increases the proportional gain, the system becomes faster, but care must be taken not make the system unstable. Once P has been set to obtain a desired fast response, the integral term is increased to stop the oscillations. The integral term reduces the steady state error, but increases overshoot. Some amount of overshoot is always necessary for a fast system so that it could respond to changes immediately. The integral term is tweaked to achieve a minimal steady state error. Once the P and I have been set to get the desired fast control system with minimal steady state error, the derivative term is increased until the loop is acceptably quick to its set point. Increasing derivative term decreases overshoot and yields higher gain with stability but would cause the system to be highly sensitive to noise. Often times, engineers need to tradeoff one characteristic of a control system for another to better meet their requirements.

The Ziegler-Nichols method is another popular method of tuning a PID controller. It is very similar to the trial and error method wherein I and D are set to zero and P is increased until the loop starts to oscillate. Once oscillation starts, the critical gain K$_c$ and the period of oscillations P$_c$ are noted. The P, I and D are then adjusted as per the tabular column shown below.

Table: Ziegler-Nichols tuning, using the oscillation method.

Control	P	Ti	Td
P	0.5Kc	-	-
PI	0.45Kc	Pc/1.2	-
PID	0.60Kc	0.5Pc	Pc/8

NI LabVIEW and PID

LabVIEW PID toolset features a wide array of VIs that greatly help in the design of a PID based control system. Control output range limiting, integrator anti-windup and bumpless controller output for PID gain changes are some of the salient features of the PID VI. The PID Advanced VI includes all the features of the PID VI along with non-linear integral action, two degree of freedom control and error-squared control.

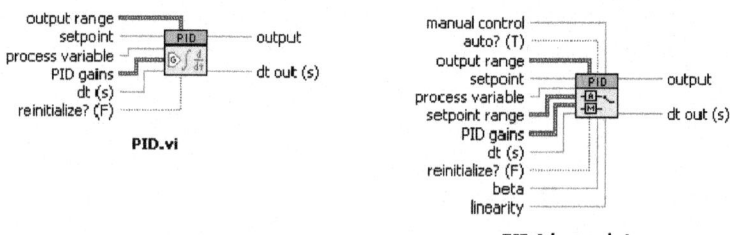

Fig. : VIs from the PID controls palette of LabVIEW.

PID palette also features some advanced VIs like the PID Autotuning VI and the PID Gain Schedule VI. The PID Autotuning VI helps in refining the PID parameters of a control system. Once an educated guess about the values of P, I and D have been made, the PID Autotuning VI helps in refining the PID parameters to obtain better response from the control system.

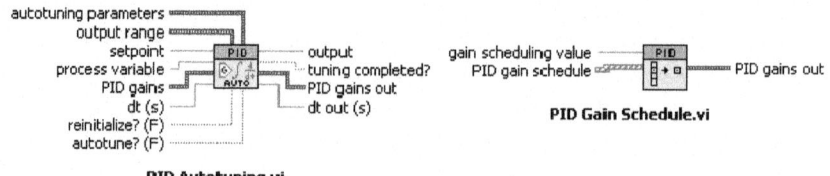

Fig. : Advanced VIs from the PID controls palette of LabVIEW.

The reliability of the controls system is greatly improved by using the LabVIEW Real Time module running on a real time target. National Instruments provides the new M Series Data Acquisition boards which provide higher accuracy and better performance than an average control system.

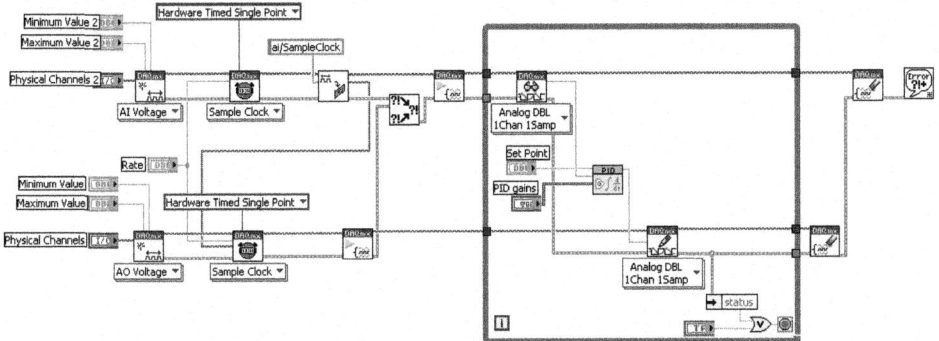

Fig. : A typical LabVIEW VI showing PID control with a plug-in
NI data acquisition device.

The tight integration of these M Series boards with LabVIEW minimizes the development time involved and greatly increases the productivity of any engineer. Figure shows a typical VI in LabVIEW showing PID control using NI-DAQmx API of M series devices.

TUNING PID LOOPS FOR LEVEL CONTROL

One-in-four control loops are regulating level, but techniques for tuning PID controllers in these integrating processes are not widely understood.

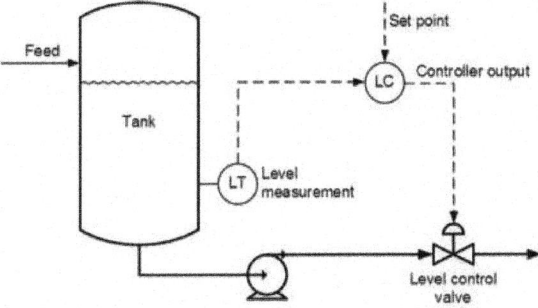

Since the first two PID controller tuning methods were published in 1942 by J. G. Ziegler and N. B. Nichols, more than 100 additional tuning rules have been developed for self-regulating control loops (*e.g.*, flow, temperature, pressure). In contrast, fewer than 10 tuning methods have been developed for integrating (*e.g.*, level) process types, though roughly one-in-four industrial PID loops controls liquid level.

The original Ziegler-Nichols tuning methods aimed for a super-fast response capability, which was achieved at the expense of control loop stability. However, a slight modification of these tuning rules improves loop stability while still maintaining a fast response to setpoint changes and disturbances. As most process experts will agree, stability is generally more important than speed.

Applicable process types

This modified Ziegler-Nichols tuning method is intended for use with integrating processes, and level control loops are the most common example.

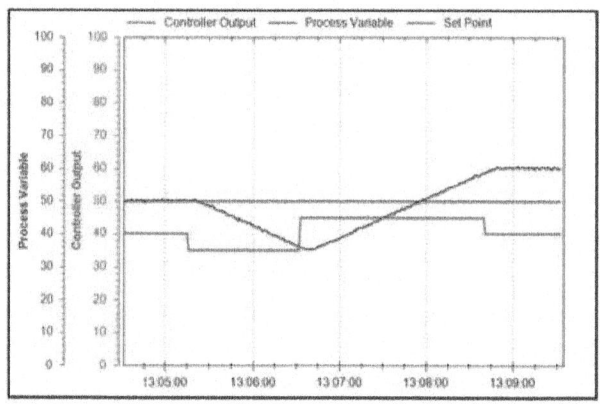

Unlike a self-regulating process, an integrating process will stabilize at only one controller output, which has to be at the point of equilibrium. If the controller output is set to a different value, the process will increase or decrease indefinitely at a steady slope.

The modified Ziegler-Nichols tuning rules presented here are designed for use on a non-interactive controller algorithm with its integral time set in minutes. Dataforth's MAQ20 industrial data acquisition and control system uses this approach as do other controllers from a variety of manufacturers.

Procedure

To apply these tuning rules to an integrating process, follow these steps. The process variable and controller output must be time-trended so that measurements can be taken from them.

Step 1. Do a Step Test

a) Make sure, as far as possible, that the uncontrolled flow in and out of the vessel is as constant as possible.

b) Put the controller in manual control mode.

c) Wait for a steady slope in the level. If the level is very volatile, wait long enough to be able to confidently draw a straight line though the general slope of the level.

d) Make a step change in the controller output. Try to make the step change 5% to 10% in size, if the process can tolerate it.

e) Wait for the level to change its slope into a new direction. If the level is volatile, wait long enough to be able to confidently draw a straight line though the general slope of the level.

   f)   Restore the level to an acceptable operating point and place the controller back into automatic control mode.

Step 2. Determine Process Characteristics

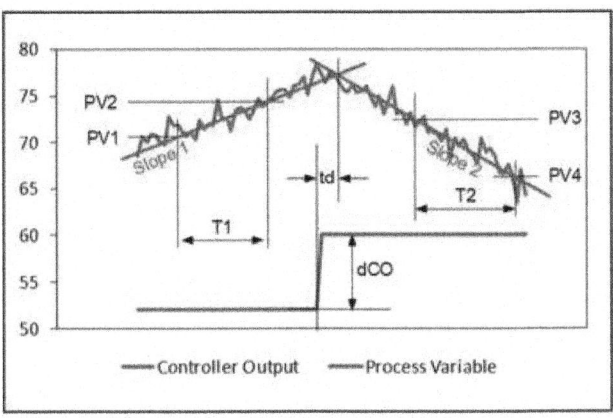

Based on the example:

a)   Draw a line (Slope 1) through the initial slope, and extend it to the right.

b)   Draw a line (Slope 2) through the final slope, and extend it to the left to intersect Slope 1.

c)   Measure the time between the beginning of the change in controller output and the intersection between Slope 1 and Slope 2. This is the process dead time (t_d), the first parameter required for tuning the controller.

d)   If t_d was measured in seconds, divide it by 60 to convert it to minutes. As mentioned earlier, the calculations here are based on the integral time in minutes, so all time measurements should be in minutes.

e)   Pick any two points (PV1 and PV2) on Slope 1, located conveniently far from each other to make accurate measurements.

f)   Pick any two points (PV3 and PV4) on Slope 2, located conveniently far from each other to make accurate measurements.

g)   Calculate the difference in the two slopes (DS) as follows:

$$DS = (PV4 - PV3) / T2 - (PV2 - PV1) / T1$$

Note: If T1 and T2 measurements were made in seconds, divide them by 60 to convert them to minutes.

h) If the PV is not ranged 0%-100%, convert DS to a percentage of the range as follows:

$$DS\% = 100 \times DS / (PV \text{ range max} - PV \text{ range min})$$

i)   Calculate the process integration rate (r_i), which is the second parameter needed for tuning the controller:

$$r_i = DS \text{ [in \%]} / dCO \text{ [in \%]}$$

Step 3. Repeat

Perform steps 1 and 2 at least three more times to obtain good average values for the process characteristics t_d and r_i.

Step 4. Calculate Tuning Constants

Using the equations below, calculate your tuning constants. Both PI and PID calculations are provided since some users will select the former based on the slow-moving nature of many level applications.

For PI Control

Controller Gain, Kc = $0.45 / (r_i \times t_d)$

Integral Time, Ti = $6.67 \times t_d$

Derivative Time, Td = 0

For PID Control

Controller Gain, Kc = $0.75 / (r_i \times t_d)$

Integral Time, Ti = $5 \times t_d$

Derivative Time, Td = $0.4 \times t_d$

Note that these tuning equations look different from the commonly published Ziegler-Nichols equations. The first reason is that Kc has been reduced and Ti increased by a factor of two, to make the loop more stable and less oscillatory. The second reason is that the Ziegler-Nichols equations for PID control target an interactive controller algorithm, while this approach is designed for a non-interactive algorithm such as is used in the Dataforth MAQ20 and others. (If you are using a different controller, make sure you find out which approach it uses.) The PID equations above have been adjusted to compensate for the difference.

Step 5. Enter the Values

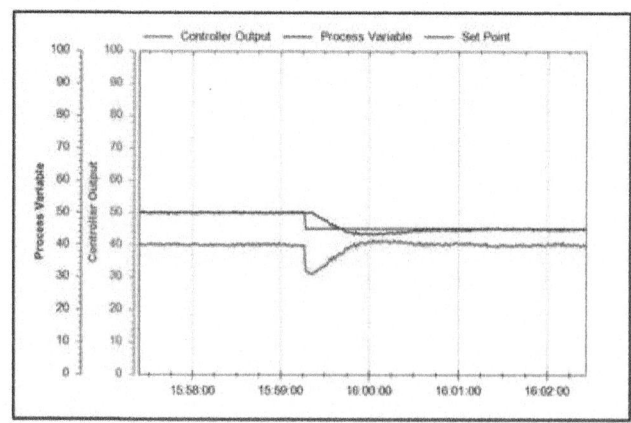

Key your calculated values into the controller, making sure the algorithm is set to non-interactive, and put the controller back into automatic mode.

Step 6. Test and Tune Your Work

Change the setpoint to test the new values and see how it responds. It might still need some additional fine-tuning to look. For integrating processes, Kc and Ti need to be adjusted simultaneously and in opposite directions. For example, to slow down the control loop, use Kc/2 and Ti × 2.

With just a few modifications to the original Ziegler-Nichols tuning approach, these rules can be used to tune level control loops for both stability and fast response to setpoint changes and disturbances.

OPTIMAL TUNING OF PID CONTROLLERS

The measure of the quality of the transient response of a PID controlled system can be performed by calculating an integral performance index. The best controller is one that has the minimum performance index. When this performance index is a minimum for a specified input, the system performance is said to be optimal. When the input signal is specified the quadratic performance index J_{ISE} can be calculated for a given plant transfer function as a function of the tuning parameters, $e.g.$ K_C, T_I, T_D and T_V.

The mathematical calculation of this performance index for given values of the tuning parameters is simple. But getting the optimal parameters is a non-trivial task. Though computerised optimisation algorithms are available to calculate the optimal parameter setting, for the case of quadratic performance indices a mathematical analysis is possible.

In the following the command and disturbance behaviour of a control system with a real PID controller and a plant with the transfer function

$$G_P(s) = \frac{K_P}{(1+Ts)^4}.$$

will be investigated. The response of the control error to step changes $w(t) = w_o \sigma(t)$ in the command input and $z'(t) = z_o \sigma(t)$ in the plant input is

$$E(s) = \frac{w_o - z_o G_C}{1 + G_C G_P} \frac{1}{s}.$$

For the plant and the real PID controller one obtains

$$E(s) = \frac{w_o T_I (1+Ts)^1 (1+T_V s) - z_o K_P T_I (1+T_V s)}{T_I (1+Ts)^1 (1+T_V s) + K_C K_P [1 + (T_I + T_V)s + (T_D + T_V)s^2]},$$

which is in the form of for $k = K_C K_P$.

Applying the analysis to the J_{ISE} performance index one gets,

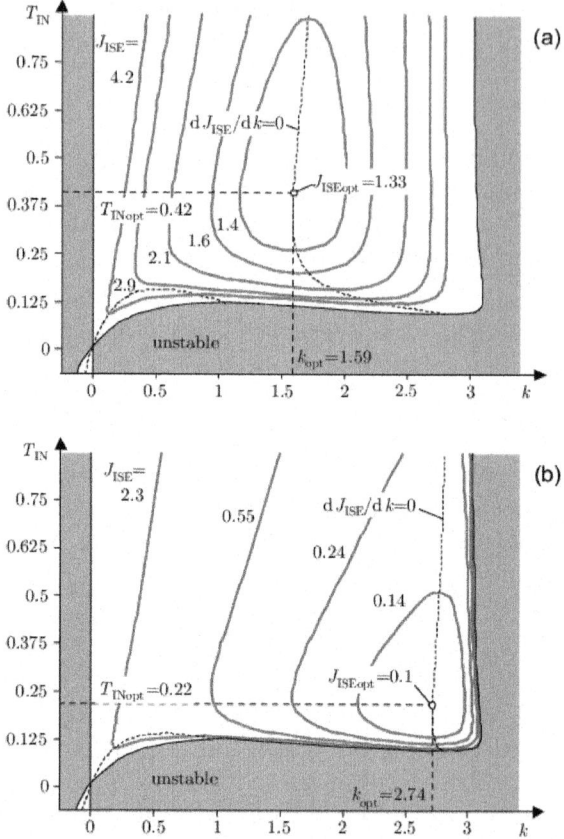

Fig. : Stability and performance diagram for step changes (a) in the command input $(w_o = 1, z_o = 0)$ and (b) in the plant input $(z_o = 1, w_o = 0)$

separately for the command and disturbance inputs. The integral action time constant is normalised by $T_{IN} = \dfrac{T_I}{4T}$. These diagrams are shown for the optimal value $T_D = T_{Dopt} = \dfrac{s}{3}T$ of the derivative action time constant. The filter time constant is $T_V = 0.1\ T_D$. The diagrams show a rather rectangular stability area that makes tuning of K_C and T_I for a fixed T_D easy from the stability point of view. But the performance characteristics are quite different. The optimal parameters for the two cases differ by about a factor of two. Therefore, an optimal tuned controller is in general never optimally tuned for command and disturbance inputs.

Advantages and Disadvantages of the Different Types of Controllers

In the following the disturbance behaviour is investigated using the controllers. Their parameters are tuned optimally according to the performance index J_{ISE}. The plant is given by Equation. The different types of controller the responses to a step disturbance $z_0\sigma(t)$ of the controlled variable y, which is normalised by $K_p z_0$. These curves indicate that because $w(t) \equiv 0$ the relation $e(t) = -y(t)$ is valid.

For discussing these curves the term *settling time* $t_{3\%}$ is used, which is related to the steady state of the uncontrolled case

$$y_{\infty, \text{without}} = K_p z_0.$$

In addition, the different cases should be compared with respect to the normalised maximum overshoot $M_P/(K_p z_0)$.

The different cases are discussed below:

a) The *P controller* shows a relatively high maximum overshoot $M_P/(K_p z_0)$, a long settling time $t_{3\%}$ as well as a steady-state error e_∞.

b) The *I controller* has a higher maximum overshoot than the P controller due to the slowly starting I behaviour, but no steady-state error.

c) The *PI controller* fuses the properties of the P and I controllers. It shows a maximum overshoot and settling time similar to the P controller but no steady-state error.

d) The real *PD controller* according to Equation with $T_V = T_D/10$ has a smaller maximum overshoot due to the 'faster' D action compared with the controller types mentioned under a) to c). Also in this case a steady-state error is visible, which is smaller than in the case of the P controller. This is because the PD controller generally is tuned to have a larger gain K_C due to the positive phase shift of the D action. For the results the gain for the P controller is $K_C = 2.68$ and for the PD controller $K_C = 4.74$. The plant has a gain of $K_p = 1$.

e) The PID controller according to Equation with $T_V = T_D/10$ fuses the properties of a PI and PD controller. It shows a smaller maximum overshoot than the PD controller and has no steady state error due to the I action.

The qualitative concepts of this example are also relevant to other type of plants with delayed proportional behaviour. This discussion has given some first insights into the static and dynamic behaviour of control loops.

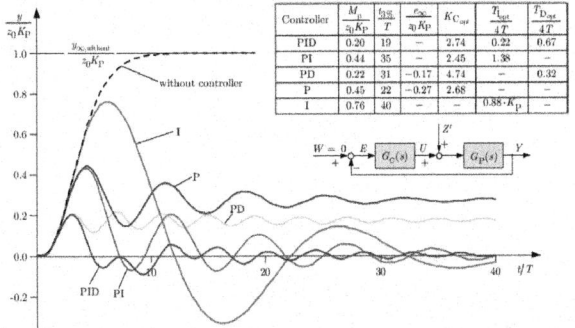

Controller	$\dfrac{M_p}{z_0 K_P}$	$\dfrac{t_{3\%}}{T}$	$\dfrac{e_\infty}{z_0 K_P}$	$K_{C_{at}}$	$\dfrac{T_{I_{at}}}{4T}$	$\dfrac{T_{D_{at}}}{4T}$
PID	0.20	19	–	2.74	0.22	0.67
PI	0.44	35	–	2.45	1.38	–
PD	0.22	31	–0.17	4.74	–	0.32
P	0.45	22	–0.27	2.68	–	–
I	0.76	40	–	–	$0.88 \cdot K_p$	–

Fig. : Behaviour of the normalised controlled variable $y/(z_0 K_p)$ for step disturbance $z' = z_0\,\sigma(t)$ at the input to the plant $[G_p(s) = K_p/(1 + Ts)^4;\ K_p = 1]$ for different types of controllers.

Empirical Tuning Rules According to Ziegler and Nichols

Many industrial processes show step responses with pure aperiodic behaviour according. This S-shape curve is characteristic of many high-order systems and such plant transfer functions may be approximated by the mathematical model

$$G_P(s) = \frac{K_P}{1+Ts}e^{-T_u s},$$

which contains a 1st-order delay element and a dead time. The approximation by a $PT_1\ T_t$ element.

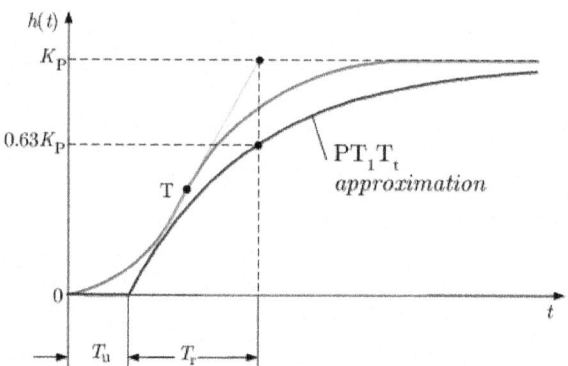

Fig. : Describing the step response of a process by the three characteristic values K_P (gain of the plant), T_t (rise time) and T_u (delay time).

Here the step response is characterised by constructing the tangent at the turning point T with the following three values: K_p (gain of the plant), T_t (rise time) and T_u (delay time). Then a rough approximation according to Equation is to set $T_t = T_u$ and $T = T_t$.

For a plant of the type described above a lot of tuning rules for standard controllers have been developed. These have been mostly developed empirically from simulation studies. The most famous empirical tuning rules are those of *Ziegler* and *Nichols*. These tuning rules have been derived to provide step responses for the closed loop, where the response shows a decrease of the amplitude of approx. 25% per period. For the application of these rules according to Ziegler and Nichols two different approaches can be used:

a) *Method of the stability margin(I):* Here, the following steps are used:

1. The controller is switched to pure P action.

2. The gain K_C of the P controller is continuously increased until the closed loop shows permanent oscillations. The value of the gain K_C at this state is denoted as the critical controller gain K_{Ccrit}.

3. The length of period T_{crit} (critical period) of the oscillations is measured.

4. From K_{Ccrit} and T_{crit} one determines the controller tuning values K_C, T_I and T_D using the formulas given in Table.

b) *Method of the step response (II):* In the case of an industrial plant it is often not possible, suitable or allowed to drive the plant into permanent oscillations for determining K_{Ccrit} and T_{crit}. Measuring the step response of the plant does not generally cause difficulties. Therefore, in many cases the second form of the Ziegler-Nichols approach is more expedient. The rules are based directly on the slope K_p/T_r of the tangent at the turning point and on the delay T_u of the step response. One has to observe that the measurement of the step response needs only to be taken at the turning point T, as the slope of the tangent already describes the ratio K_p/T_r. Using the measured data T_u and K_p/T_r as well as the formula given in Table below the controller tuning parameters can be determined by simple calculations.

Table: Controller tuning parameters according to Ziegler and Nichols

	Type of controller	Controller parameters		
		K_C	T_1	T_D
Method I	P	$0.5\, K_{Ccrit}$	-	-
	PI	$0.45\, K_{Ccrit}$	$0.85\, T_{crit}$	-
	PID	$0.6\, K_{Ccrit}$	$0.5\, T_{crit}$	$0.12\, T_{crit}$
Method II	P	$\dfrac{1}{K_P}\dfrac{T_r}{T_u}$	-	-
	PI	$\dfrac{0.9}{K_P}\dfrac{T_r}{T_u}$	$3.33\, T_u$	-
	PID	$\dfrac{1.2}{K_P}\dfrac{T_r}{T_u}$	$2\, T_u$	$0.5\, T_u$

ZIEGLER-NICHOLS OPEN-LOOP TUNING RULES

- The Ultimate Cycling method, and
- The Process Reaction-Curve method, often called the Ziegler-Nichols Open-Loop tuning method.

Quarter-Amplitude Damping

The Ziegler-Nichols tuning methods aim for a quarter-amplitude damping response. Although the quarter-amplitude damping type of tuning provides very fast rejection of disturbances, it makes the loop very oscillatory, often causing interactions with similarly-tuned loops. Quarter-amplitude damping-type tuning also leaves the loop vulnerable to going unstable if the process gain or dead time increases.

The easy fix for both problems is to reduce the controller gain by half. However, if the control objective for the loop you are tuning is to have a very stable, robust control loop that absorbs disturbances, rather use the Lambda tuning rules.

Designed for the Interactive Controller Algorithm

There are three types of PID controller algorithms: Interactive, Noninteractive, and Parallel. The Ziegler-Nichols tuning rules were designed for controllers with the interactive controller algorithm. If you are not using the derivative control mode (*i.e.* using P or PI control), the rules will also work for the noninteractive algorithm. However, if you plan to use derivative (*i.e.* PID control) and have a noninteractive controller, or if your controller has a parallel algorithm, you should convert the calculated tuning settings to work on your controller.

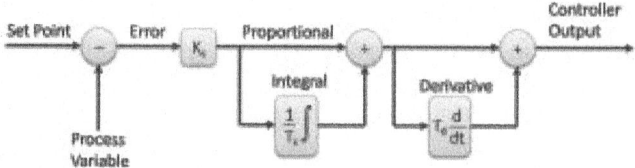

Use of Integral Time

The original Ziegler-Nichols tuning rules were designed for controllers using reset rate (integral gain in repeats per minute) and not integral time (in minutes or seconds). However, virtually all the modern texts on process control use integral time. If your controller uses integral gain or reset rate, you'll have to invert the calculated integral time (use $1/Ti$).

Limited Range of Process Dynamics

The Ziegler-Nichols tuning rules work well on processes of which the time constant is at least two times as long as the dead time. For example temperature and gas pressure. They work moderately poorly on flow loops and liquid pressure loops where the dead time and time constant are about equal in length. And they work very poorly on dead-time dominant processes. The Cohen-Coon tuning rules work better on a wider range of processes.

Slight Modification for Self-Regulating Processes

The tuning method described below is actually a widely-used modification of the published Ziegler-Nichols Process Reaction Curve method. The reaction-curve method was designed for use on integrating and self-regulating processes. The modified method works only on self-regulating processes, but then more accurately so. The modified method is so popular that few people know about the original reaction-curve method. I will describe the reaction-curve method in a future post, because it works very well for integrating processes.

Tuning Procedure

Assuming the control loop is linear and the final control element is in good working order, you can continue with tuning the controller. The Ziegler-Nichols open-loop tuning rules use three process characteristics: process gain, dead time, and time constant. These are determined by doing a step test and analyzing the results.

1. Place the controller in manual and wait for the process to settle out.
2. Make a step change of a few percent in the controller output (CO) and wait for the process variable (PV) to settle out at a new value. The size of this step

should be large enough that the process variable moves well clear of the process noise/disturbance level. A total movement of five times the noise/disturbances on the process variable should be sufficient.

3. Convert the total change obtained in PV to a percentage of the span of the measuring device.

4. Calculate the process gain (gp) as follows:

    o   gp = change in PV [in %] / change in CO [in %]

5. Find the maximum slope on the PV response curve. This will be at the inflection point(where the PV stops curving upward and begins curving downward). Draw a line tangential to the PV response curve through the point of inflection. Extend this line to intersect with the original level of the PV (before the step-change in CO). Take note of the time value at this intersection.

6. Measure the dead time (td) as follows:

    o   td = time difference between the step-change in CO and the intersection described above.

7. Calculate the value of the PV at 63% (0.63) of its total change. On the PV reaction curve, find the time value at which the PV reaches this level.

8. Measure the time constant (Greek symbol tau) as follows:

    o   tau = time difference between intersection at the end of dead time, and the PV reaching 63% of its total change.

9. Convert your measurements of dead time and time constant to the same time-units your controller's integral mode uses. *E.g.* if your controller's integral time is in minutes, use minutes for these measurements.

10. Do two or three more step tests and calculate process gain, dead time, and time constant for each test to obtain a good average of the process characteristics. If you get vastly different numbers every time, do even more step tests until you have a few step tests that produce similar values. Use the average of those values.

11. Calculate settings for Controller Gain (Kc), Integral Time (Ti), and Derivative Time (Td), using the Ziegler-Nichols tuning rules below. Note that these rules produce a quarter-amplitude damping response and the calculated controller gain values should be divided by two.

    o   For P control: Kc = tau / (gp * td)

    o   For PI control: Kc = 0.9 * tau / (gp * td); Ti = 3.33 * td

    o   For PID control: Kc = 1.2 * tau / (gp * td); Ti = 2 * td; Td = 0.5 * td

12. IMPORTANT: If you have not already done so, divide the calculated controller gain (Kc) by two to reduce overshoot and improve stability.

13. Compare the newly calculated controller settings with the ones in the controller, and ensure that any large differences in numbers are expected and justifiable.

14. Make note of the previous controller settings, the new settings, and the date and time of change.

15. Implement and test the new controller settings. Ensure the response is in line with the overall control objective of the loop.

16. Leave the previous controller settings with the operator in case he/she wants to revert back to them and cannot find you to do it. If the new settings don't work, you have probably missed something in one or more of the previous steps.

17. Monitor the controller's performance periodically for a few days after tuning to verify improved operation under different process conditions.

Ziegler - Nichols PID Tuning

In order to address an FAQ we present here a brief overview of the Ziegler-Nichols (short: Z-N) tuning methods.

Ziegler and Nichols have developed PID tuning methods back in the early fourties based on open loop tests (less known than for example the Cohen-Coon formulas) and also based on a closed loop test, which is maybe their most widely known achievement.

The *open loop method* allows to calculate PID parameters from the process parameters. The procedure:

- Step 1: Make an open loop plant test (*e.g.* a step test)
- Step 2: Determine the process parameters: Process gain, deadtime, time constant.
- inflection point and measure L and T as shown. By the way: Today we have better and easier methods).
- Step 3: Calculate the parameters according to the following formulas:

K = time constant / (process gain * deadtime)

PI: Proportional gain = 0.9 * K, integral time = 3.3 * deadtime

PID: Proportional gain = 1.2 * K, integral time = 2 * deadtime, derivative time = 0.5 * deadtime

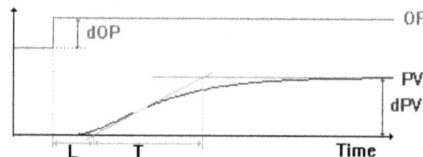

Process gain = dPV / dOP, deadtime = L, time constant = T

The *closed loop method* prescribes the following procedure:

- Step 1: Disable any D and I action of the controller (--> pure P-controller)
- Step 2: Make a setpoint step test and observe the response
- Step 3: Repeat the SP test with increased / decreased controller gain until a stable oscillation is achieved. This gain is called the "ultimate gain" Ku.

- Step 4: Read the oscillation period Pu.
- Step 5: Calculate the parameters according to the following formulas:

PI: Proportional gain = 0.45 * Ku, integral time =Pu / 1.2

PID: Proportional gain = 0.6 * Ku, integral time =Pu / 2, derivative time = Tu / 8

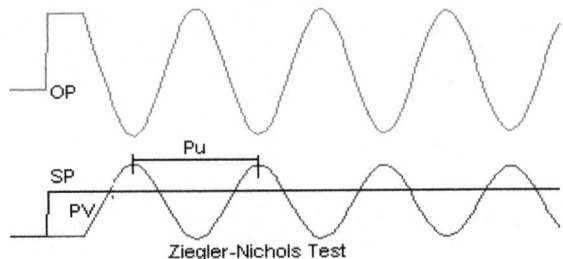

Ziegler-Nichols Test

Characterization:

- Both methods give a good starting point but require further fine-tuning.
- The open loop method is based on a measurement range of 0-100 and continuous control. This requires adjustments for other measurement ranges and for the control interval in digital systems (the method was developed in the times when only analog controllers existed).
- The closed loop methods does not require adjustments, a big advantage, since both process and controller are part of the test, but suffers from one major disadvantage: Bringing the loop into stable, sustained oscillation is simply out of the question for industrial processes.
- Both methods do not distinguish between setpoint and load tuning and are for self- regulating processes only, not for integrating processes like liquid level.

Today's Technology

In our tools ACT-TOP and TOPAS a refined Z-N methods is used:

A) **Closed loop test:** The basic approach is the same as with the original Z-N but with one major improvement: Stable oscillation is not required any more. In addition, the tools not only calculate the PID constants but also the process parameters- just from one setpoint test - and just by taking 5 data points!

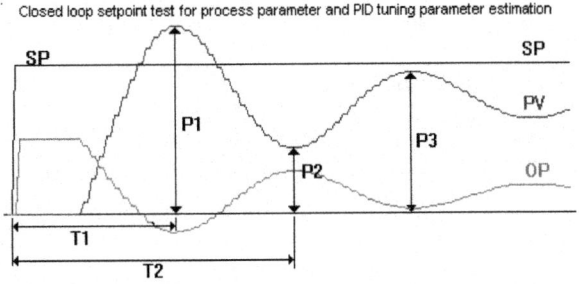

Closed loop setpoint test for process parameter and PID tuning parameter estimation

B) **Open loop test:** ACT-TOP and TOPAS provide several methods to calculate the process parameters from a step or a relay test. Once the process parameters are known you can calculate refined PID constants (less overshoot, smoother approach than Z-N) for both setpoint tuning and load tuning (P-action on error or PV) using ACT's proprietary methods. And you can calculate tuning constants for tight and average level control - to your specification.

For the more curious: Since the process parameters are known you can also compare (and measure the performance) of the PID with model based control right away - without any prior knowledge.

COHEN-COON TUNING RULES

Based on the number of Google searches in 2010, the Cohen-Coon tuning rules are second in popularity only to the Ziegler-Nichols tuning rules. Cohen and Coon published their tuning method in 1953, eleven years after Ziegler and Nichols published theirs.

More Flexible than Ziegler-Nichols

The Cohen-Coon tuning rules are suited to a wider variety of processes than the Ziegler-Nichols tuning rules. The Ziegler-Nichols rules work well only on processes where the dead time is less than half the length of the time constant.

The Cohen-Coon tuning rules work well on processes where the dead time is less than two times the length of the time constant (and you can stretch this even further if required).

Cohen-Coon provides one of the few sets of tuning rules that has rules for PD controllers – should you ever need this.

Quarter-Amplitude Damping

Like the Ziegler-Nichols tuning rules, the Cohen-Coon rules aim for a quarter-amplitude damping response. Although quarter-amplitude damping-type of tuning provides very fast disturbance rejection, it tends to be very oscillatory and frequently interacts with similarly-tuned loops. Quarter-amplitude damping-type tuning also leaves the loop vulnerable to going unstable if the process gain or dead time doubles in value. However, the easy fix for both problems is to reduce the controller gain by half. *E.g.* if the rule recommends using a controller gain of 1.8, use only 0.9. This will prevent the loop from oscillating around its set point as described above, and will provide an acceptable stability margin.

Target PID Controller Algorithm

There are three types of PID controller algorithms: Interactive, Noninteractive, and Parallel. The Cohen-Coon tuning rules were designed for controllers with the noninteractive controller algorithm. If you are not using the derivative control mode (*i.e.* using P, PI, of PD control), the rules will also work for the interactive algorithm. However, if you are using derivative (*i.e.* PID control) on an interactive controller, or if your controller has a parallel algorithm, you should convert the calculated tuning settings to work on your controller.

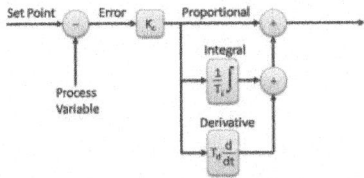

Fig. : Noninteractive Controller Structure.

A Note on Integral Time

The original Cohen-Coon paper expressed the tuning constant for the integral control mode in terms of reset rate (or integral gain) in repeats per minute. Virtually all the modern texts on process control use integral time, and so do most control systems (DCS & PLC). Also, this blog generically uses integral time and not integral gain. Therefore, the tuning rules below use integral time (the reciprocal of what Cohen-Coon used). If your controller uses integral gain or reset rate, you'll have to invert the calculated integral time (use $1/Ti$).

Also, if your controller's integral time unit is in minutes, you must make your measurements of dead time and time constant in minutes. Likewise if your controller uses seconds, make your measurements in seconds.

When to use the Cohen-Coon Tuning Rules

The Cohen-Coon tuning rules are suitable for use on self-regulating processes if the control objective is having a fast response, but I recommend you divide the calculated controller gain by two, as described above.

If the control objective is to have a very stable, robust control loop that absorbs disturbances, rather use the Lambda tuning rules.

Tuning Procedure

Assuming the control loop is linear and the final control element is in good working order, you can continue with tuning the controller. The Cohen-Coon tuning rules use three process characteristics: process gain, dead time, and time constant. These are determined by doing a step test and analyzing the results.

1. Place the controller in manual and wait for the process to settle out.

2. Make a step change of a few percent in the controller output (CO) and wait for the process variable (PV) to settle out at a new value. The size of this step should be large enough that the process variable moves well clear of the process noise/disturbance level. A total movement of five times the noise/disturbances on the process variable should be sufficient.

3. Convert the total change obtained in PV to a percentage of the span of the measuring device.

4. Calculate the process gain (gp) as follows:
   o gp = change in PV [in %] / change in CO [in %]

5. Find the maximum slope on the PV response curve. This will be at the inflection point(where the PV stops curving upward and begins curving downward). Draw a line tangential to the PV response curve through the point of inflection. Extend this line to intersect with the original level of the PV (before the step change in CO). Take note of the time value at this intersection.

6. Measure the dead time (td) as follows:

o td = time difference between the change in CO and the intersection of the tangential line and the original PV level.

7. Calculate the value of the PV at 63% of its total change. On the PV reaction curve, find the time value at which the PV reaches this level.

8. Measure the time constant (Greek symbol tau) as follows:
   o tau = time difference between intersection at the end of dead time, and the PV reaching 63% of its total change.

9. Convert your measurements of dead time and time constant to the same time-units your controller's integral mode uses. *E.g.* if your controller's integral time is in minutes, use minutes for this measurement.

10. Do two or three more step tests and calculate process gain, dead time, and time constant for each test to obtain a good average of the process characteristics. If you get vastly different numbers every time, do even more step tests until you have a few step tests that produce similar values. Use the average of those values.

11. Calculate new tuning settings using the Cohen-Coon tuning rules below. Note that these rules produce a quarter-amplitude damping response.

12. Divide the calculated controller gain by two to reduce oscillations and improve loop stability.

13. Compare the newly calculated controller settings with the ones in the controller, and ensure that any large differences in numbers are expected and justifiable.

14. Make note of the previous controller settings, the new settings, and the date and time of change.

15. Implement and test the new controller settings. Ensure the response is in line with the overall control objective of the loop.

16. Leave the previous controller settings with the operator in case he/she wants to revert back to them and cannot find you to do it. If the new settings don't work, you have probably missed something in one or more of the previous steps.

17. Monitor the controller's performance periodically for a few days after tuning to verify improved operation under different process conditions.

FUNDAMENTALS OF LAMBDA TUNING

Lambda tuning is a form of internal model control (IMC) that endows a proportional-integral (PI) controller with the ability to generate smooth, non-oscillatory control efforts when responding to changes in the setpoint. Its name derives from the Greek letter lambda (λ), which designates a user-specified performance parameter that dictates how long the controller is allowed to spend on the task of moving the process variable from point A to point B.

Like its more famous cousin, Ziegler-Nichols tuning, lambda tuning involves a set of formulas or tuning rules that dictate the values of the PI parameters required to achieve the desired controller performance. The first step in applying them is to determine how much and how fast the process responds to the controller's efforts.

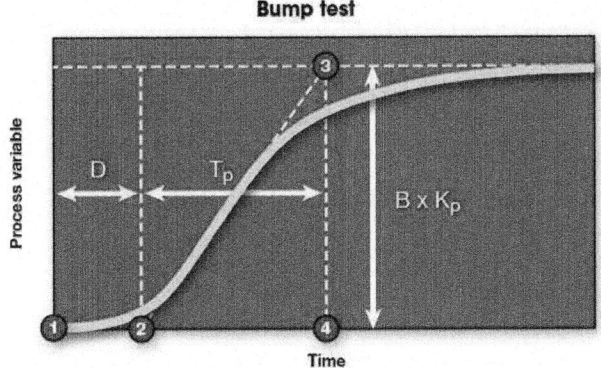

Fig. : Bump test.

This 9-step test, also known as an open-loop reaction curve test or step test, gives a PI controller everything it needs to know about the behaviour of a non-oscillatory process in order to control it:

1. Turn off the controller by switching it to manual mode.

2. Wait until the process variable settles out to a steady-state value.

3. Manually "bump" or "step" the process by forcing the control effort abruptly upwards by B% — whatever it takes to make the process variable move appreciably but not excessively.

4. Record the process variable's reaction or step response on a trend chart as above, starting at the time when the bump was applied (step 1) and ending when the process variable settles out again.

5. Draw an ascending line tangent to the steepest part of the process variable's trend line.

6. Draw horizontal lines through the process variable's initial and final values.

7. Mark where the two horizontal lines intersect the ascending line at points 2 and 3.

8. Record the deadtime D from point 1 to point 2 and the process time constant T_p from point 2 to point 4.

9. Record the change in the process variable from point 3 to point 4 then divide that by B to get the process gain K_p.

Once the process behaviour has been characterized in terms of the process parameters — the process gain Kp, the process time constant Tp, and the deadtime D — tuning the controller is simple. Just plug those values and the user's choice of λ into the formulas shown in the "lambda tuning rules" sidebar to get the required values for the PI parameters Kc and Ti.

Consider a process with an open-loop gain Kp, a time constant Tp, and a deadtime D being driven by the control effort or *control output* CO(t) from a PI controller given by

$$CO(t) = K_c \left[e(t) + \frac{1}{T_i} \int e(t)dt \right]$$

where

$$e(t) = SP(t) - PV(t)$$

is the error at time t between the process variable PV(t) and the setpoint SP(t). The rules for lambda tuning call for

$$K_c = \frac{T_p}{K_p(\lambda + D)}$$

and

$$T_i = T_p$$

in order to obtain a closed-loop system with a non-oscillatory setpoint response that will settle out in approximately 4λ seconds.

Note that these tuning rules require the user to specify only one performance parameter: λ. This not only simplifies the calculation of K_c and T_i, but it also allows the user to select the controller's desired performance in terms of a physically meaningful quantity — the time allowed to complete a setpoint change — as opposed to the less intuitive concepts of proportional band and reset time.

Closed-loop Performance

A PI controller thus tuned will, theoretically, complete a setpoint change in about 4λ seconds when operating in closed-loop mode, and it will do so without overshoot. That is, it will drive the process variable towards the setpoint gradually enough to guarantee that the error between them will continue to diminish steadily.

This overdamping feature can be especially useful in applications where the process variable must be maintained near some limiting value that the process variable must not cross. The controller will never accidentally violate such a constraint because it will never drive the process variable past the setpoint. Nor will a lambda-tuned controller ever cause unstable oscillations in the process variable because it will never need to reverse course after a setpoint change. The process variable will always proceed steadily upward or steadily downward until the new setpoint is reached.

Overdamping also helps ensure consistency, which is why lambda tuning has become particularly popular for paper-making operations where fluctuations in certain process variables can cause visible irregularities in the finished product. The absence of overshoot also prevents the interacting loops in a paper-making machine from shaking the equipment to death by causing the actuators to oscillate all at once. And individual actuators (especially valves) will be subject to less wear and tear since they will never be required to reverse course unless the setpoint does.

Coordinating Multiple Loops

Furthermore, since lambda tuning allows a PI controller to achieve its objective over a user-specified interval, it can be used to synchronize all of the controllers in a multi-loop operation so that the process variables will all move at roughly the same rate. This too contributes to uniformity in the paper-making process. It also helps maintain a constant ratio of ingredients in a blending operation.

Conversely, when certain interacting loops are more important than others, the most critical ones can be assigned smaller λ values to make sure that they remain out of spec for the shortest possible interval following a setpoint change. Loops that contribute less to the overall profitability of the operation and loops that have slower or less powerful actuators can be allowed to take their time with larger λ values.

Using highly disparate lambda values for two interacting loops can also help decouple them. The faster loop will see little or no effect from the slower one since the latter will appear to be more-or-less stationary during the interval that the former is completing its latest setpoint change. Conversely, the faster loop will have finished its work by the time the slower one gets underway. The decoupling won't be complete, but the interactions between the loops will be mitigated at least somewhat, thereby reducing the apparent loads that each would otherwise cause the other.

A less obvious advantage of lambda tuning is its robustness. Because a lambda-tuned controller is so conservative, it can withstand considerable discrepancies between the estimated and actual values of the process parameters, whether those discrepancies are due to a poorly executed bump test or a change in the process that occurs after the tuning is complete. The resulting distortions in the calculated PI parameters may well make the controller more or less conservative than if the tuning had been accomplished with perfect knowledge of the process's behaviour, but the closed-loop system is likely to remain overdamped either way.

Disadvantages

On the other hand, lambda tuning has its limits, especially when speed is of the essence. It tends to make a slow process even slower, causing the process variable to remain out of spec for a long time. Specifically, λ is generally assigned a value between Tp and 3Tp, making the closed-loop response to a setpoint change up to three times longer than the corresponding open-loop step response. An even larger value of λ is required if the dead time D is significant. In such cases, $\lambda > D$ is the practical lower limit since the controller can't be expected to react any faster than the dead time allows.

But arguably the most challenging drawback to a lambda-tuned controller is its limited ability to deal with an external load on the process. It can still bring the process variable back to the setpoint if a random load should ever upset the process variable, but it will make no effort to do so particularly quickly or efficiently. Even measuring the disturbances won't help because the lambda tuning rules make no provisions for the behaviour of the load, only the process.

The best the user can do is to set λ as small as possible in order to increase the controller's speed overall, but doing so will tend to make the controller less robust. And lambda tuning wouldn't be a particularly good choice anyway when a fast response is required since there are other tuning rules that are much more effective for time-sensitive applications.

Mathematical Challenges

There are also some subtle limitations to lambda tuning buried deep in the underlying mathematics. For one, it can't be applied to a process that is itself oscillatory. If an open-loop bump test yields a step response that fluctuates before settling out, the process cannot be completely characterized by just the three parameters K_p, T_p, and D and the controller cannot be tuned with the lambda rules, though there are several related IMC techniques that will work just fine in such cases.

The mathematics also break down when the deadtime D is especially large. The calculations required to compute K_c and T_i suffer from an approximation that becomes less and less accurate as D increases. Several alternative approaches have been proposed to improve the accuracy of that approximation, but those efforts have also generated considerable confusion — multiple sets of tuning rules all called "lambda tuning." They all achieve roughly the same closed-loop performance but look nothing alike. Some apply to a PI-only controller while others require a full PID controller equipped with derivative action.

Lambda tuning rules also take on different forms for an integrating process that has no time constant. These occur in applications such as level control where a bump from the controller (opening the inlet valve) results in a process variable (liquid level) that continues to rise without leveling off. A lambda-tuned controller can force an integrating process to reach a steady-state, but it takes longer for the process variable to settle out — about 6 λ seconds — and the process variable will overshoot the setpoint along the way.

Nonetheless, lambda tuning is relatively simple, intuitive, and bullet-proof. It will no doubt remain popular in applications where a conservative controller is required.

LAMBDA TUNING RULES

The Lambda tuning rules, sometimes also called Internal Model Control (IMC)* tuning, offer a robust alternative to tuning rules aiming for speed, like Ziegler-Nichols, Cohen-Coon, *etc.* Although the Lambda and IMC rules are derived differently, both produce the same rules for a PI controller on a self-regulating process.

While the Ziegler-Nichols and Cohen-Coon tuning rules aim for quarter-amplitude damping, the Lambda tuning rules aim for a first-order lag plus dead time response to a set point change. The Lambda tuning rules offer the following advantages:

- The process variable will not overshoot its set point after a disturbance or set point change.

- The Lambda tuning rules are much less sensitive to any errors made when determining the process dead time through step tests. This problem is common with lag-dominant processes, because it is easy to under- or over-estimate the relatively short process dead time. Ziegler-Nichols and Cohen-Coon tuning rules can give really bad results when the dead time is measured incorrectly.

- The tuning is very robust, meaning that the control loop will remain stable even if the process characteristics change dramatically from the ones used for tuning.

- A Lambda-tuned control loop absorbs a disturbance better, and passes less of it on to the rest of the process. This is a very attractive characteristic for using Lambda tuning in highly interactive processes. Control loops on paper-making machines are commonly tuned using the Lambda tuning rules to prevent the entire machine from oscillating due to process interactions and feedback control.

- The user can specify the desired response time (actually the closed loop time constant) for the control loop. This provides one tuning factor that can be used to speed up and slow down the loop response.

Unfortunately, the Lambda tuning rules have a drawback too. They set the controller's integral time equal to the process time constant. If a process has a very long time constant, the controller will consequently have a very long integral time. Long integral times make recovery from disturbances very slow.

It is up to you, the controls practitioner, to decide if the benefits of Lambda tuning outweigh the one drawback. This decision must take into account the purpose of the loop in the process, the control performance objective, the typical size of process disturbances, and the impact of deviations from set point.

Below are the Lambda tuning rules for a PI controller. Although Lambda / IMC tuning rules have also been derived for PID controllers, there is little point in using derivative control in a Lambda-tuned controller. Derivative control should be used if a fast loop response is required, and should therefore be used in conjunction with a fast tuning rule (like Cohen Coon). Lambda tuning is not appropriate for obtaining a fast loop response. If speed is the objective, use another tuning rule.

To apply the Lambda tuning rules for a self-regulating process, follow the steps below. Also, please read the paragraph in red text following the tuning equations.

Do a Step-test and Determine the Process Characteristics

a) Place the controller in manual and wait for the process to settle out.

b) Make a step change in the controller output (CO) of a few percent and wait for the process variable (PV) to settle out. The size of this step should be large enough that the PV moves well clear of the process noise/disturbance level. A total movement of five times the noise/disturbances on the process variable should be sufficient.

c) Calculate the process characteristics as follows:

Process Gain (gp)

Convert the total change in PV to a percentage of the measurement span.

gp = change in PV [in %] / change in CO [in %]

Dead Time (td)

Note: Make this measurement in the same time-units your controller's integral mode uses. *E.g.* if your controller's integral time is in minutes, use minutes for this measurement.

Find the maximum slope of the PV response curve. This will be at the point of inflection. Draw a line tangential through the PV response curve at this point. Extend this line to intersect with the original level of the PV before the step in CO. Take note of the time value at this intersection.

td = time difference between the change in CO and the intersection of the tangential line and the original PV level

Time Constant (tau)

Calculate the value of the PV at 63% of its total change. On the PV reaction curve, find the time value at which the PV reaches this level

tau = time difference between intersection at the end of dead time, and the PV reaching 63% of its total change

Note: Make this measurement in the same time-units your controller's integral mode uses. *E.g.* if your controller's integral time is in minutes, use minutes for this measurement.

d) Repeat steps b) and c) two more times to obtain good average values for the process characteristics. If you get vastly different numbers every time, do even more step tests until you have a few step tests that produced similar values. Use the average of those values.

Pick a Desired Closed Loop Time Constant (taucl) for the Control Loop

A large value for taucl will result in a slow control loop, and a small taucl value will result in a faster control loop. Generally, the value for taucl should be set between one and three times the value of tau.

Use taucl = 3 x tau to obtain a very stable control loop. If you set taucl to be shorter than tau, the advantages of Lambda tuning listed above soon disappear.

Calculate PID Controller Settings Using the Equations Below

Controller Gain (Kc)

Kc = tau/(gp x (taucl + td))

Integral Time (Ti)

Ti = tau

Derivative Time (Td)

Td = zero.

Important Notes!

- The tuning equations above are designed to work on controllers with interactive or noninteractive algorithms, but **not** controllers with parallel (independent gains) algorithms.

- The rules calculate controller gain (Kc) and not proportional band (PB). PB = 100/Kc.

- The rules assume the controller's integral setting is integral time Ti (in minutes or seconds), and not integral gain Ki (repeats per minute or repeats per second). Ki = 1/Ti.

If your controller is different from the above, simple parameter conversions will allow you to use the Lambda rules.

TUNING RULE FOR DEAD-TIME DOMINANT PROCESSES

Processes with lags or time constants (tau) longer than their dead times (td) are reasonably easy to tune. Most tuning rules work well for processes where tau > 2 td (lag dominant). The opposite is not true. Many tuning rules work very poorly when td > 2 tau (dead-time dominant).

Lag Dominant

When a process has a time constant that is much longer than the dead time, problems like overshoot and having to use high controller gains begin to appear. However, loops with long time constants still act in an intuitive way – if we add more control action we can make the process respond faster, like stepping down harder on the accelerator will get our car to the desired speed quicker.

Dead-Time Dominant

On the other side of the spectrum, when a process' dead time is significantly longer than its time constant, it behaves much less intuitively – adding more control action does not make the process respond faster. For example, if your shower water is a little cold, opening the hot water tap a lot more is not going to get you to the right temperature any quicker, and it is going to have some serious side-effects.

I once saw several operators struggle to manually control the outlet temperature of a three-pass kiln. The kiln was a dead-time dominant process and its dead time was about 10 minutes long. The operators would notice the temperature is below set point and increase the firing rate. When they see no effect, they increase the firing rate more. And then some more, and more. Finally, when changes have made their way through the dead time, the temperature overshoots its set point by a large margin. Then the operators take the same actions and make the same mistakes in the opposite direction.

Needless to say, controller tuning also becomes difficult on dead-time dominant processes.

Tuning

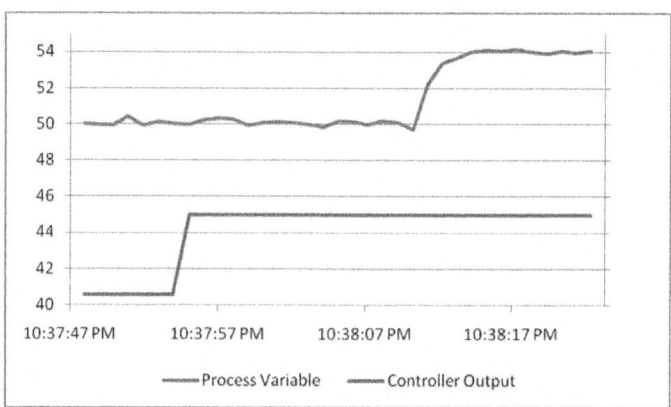

Fig. : Step response of a dead-time dominant process.

You will find that the Ziegler-Nichols tuning rules don't work well at all on a dead-time dominant process. For example, the following process characteristics were measured from the step-response of a dead-time dominant process in the previous plot:

td = 0.276 minutes

tau = 0.013 minutes

gp = 0.89

Applying the Ziegler-Nichols tuning rules to this process gives the following controller settings: Kc = 0.05; Ti = 0.92 minutes. The result is an extremely sluggish control loop.

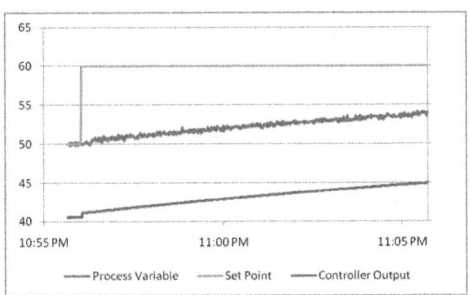

Fig. : Dead-time dominant loop tuned with the Ziegler-Nichols tuning rules.

Processes with time constants (tau) longer than their dead times (td) are reasonably easy to tune. Most tuning rules work well for processes where tau > 2 td (lag dominant). The opposite is not true. Most tuning rules work very poorly when td > 2 tau (dead-time dominant).

The Lambda tuning rules were designed for lag dominant processes and do not work all that well on dead-time dominant processes either. The Cohen-Coon tuning rules work much better than the Ziegler-Nichols rules, but they too aren't the best tuning rule when the dead time is five or ten times as long as the time constant.

So what type of tuning rule will work well for controlling dead-time dominant processes? First, we need a lag-dominant controller, to make up for the absence of lag in the process. But if we just crank up the integral term, the loop will become unstable. So, second, we have to compensate by decreasing the controller gain.

The Cohen-Coon PI tuning rules will work reasonably well up to td = 2 tau, but it becomes sluggish after that. When td > 2 tau, it is better to use the dead-time tuning rule. It is as follows:

$$Kc = 0.36 / (gp * SM)$$

$$Ti = td / 3$$

No derivative.

SM is the stability margin and can be set to a value between 1 and 4. A value of 1 is equivalent to the 1/4-amplitude damping response. It is considered unsafe – the loop is very sensitive to changes in process conditions. A value of 2 or higher is recommended. It will reduce the overshoot, eliminate unnecessary cycling, and make the loop far more robust to changes in process conditions.

Hint: measure dead time in the same units of time as your controller's integral setting. *E.g.* if your controller's Ti setting is in minutes, measure td in minutes.

Notes:

- The tuning rules above are designed to work on controllers with interactive or non-interactive algorithms, but not controllers with parallel algorithms.
- Furthermore, they will work only on controllers with a controller gain setting and not a proportional band (found on Foxboro I/A controllers, for example).
- The rules assume the controller's integral setting is in units of time (minutes or seconds), and not integral gain or rate (repeats per minute or repeats per second).

If your controller is different, parameter conversions will allow you to use these rules.

Applying the dead-time tuning rules to the process described above gives the following controller settings: Kc = 0.2; Ti = 0.092 minutes. The result is significantly better than what can be obtained with other tuning rules.

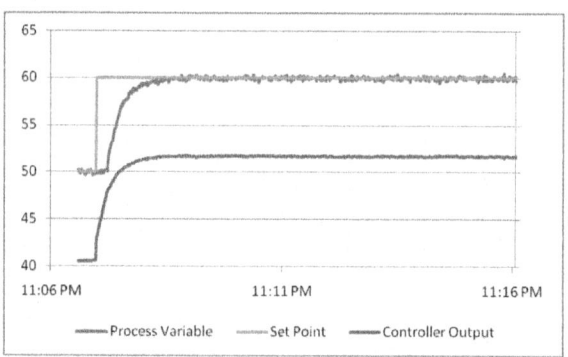

Fig. : Dead-time dominant loop tuned with the Dead-Time tuning rules.

Better loop response can be obtained with a Smith Predictor, but this is more complex to implement and very sensitive to changes in process characteristics.

A DESIGN AND TUNING RECIPE FOR INTEGRATING PROCESSES

It is best practice to follow a formal recipe when designing and tuning a PID controller. A recipe lets us move a controller into operation quickly. And perhaps most important, the performance of the controller will be superior to one tuned using an intuitive approach or trial-and-error method.

Additionally, a recipe-based approach overcomes many of the concerns that makes control projects challenging in an industrial operating environment. Specifically, a recipe approach causes less disruption to the production schedule, wastes less raw material and utilities, requires less personnel time, and generates less off-spec product.

The Recipe for Integrating Processes

Integrating (or non-self regulating) processes display counter-intuitive behaviors that make them surprisingly challenging to control. In particular, they do not naturally settle out to a steady operating level if left uncontrolled.

So while the controller design and tuning recipe is generally the same for both self regulating and integrating processes, there are important differences. Specifically, step 3 of the recipe uses a different dynamic model form and step 4 employs different tuning correlations.

Yet the design and tuning recipe maintains the familiar four step structure:

1. Establish the design level of operation (the normal or expected values for set point and major disturbances).

2. Bump the process and collect controller output (CO) to process variable (PV) dynamic process data around this design level.

3. Approximate the process data behaviour with a *first order plus dead time integrating* (FOPDT Integrating) dynamic model.

4. Use the model parameters from step 3 in rules and correlations to complete the controller design and tuning.

It is important to recognize that real processes are more complex than the simple FOPDT Integrating model form used in step 3. In spite of this, the FOPDT Integrating model succeeds in providing an approximation of process behaviour that is sufficiently accurate to yield reliable and predictable control performance when used with the rules and correlations in step 4 of the recipe.

The FOPDT Integrating Model

We recall that the familiar first order plus dead time (FOPDT) dynamic model used to approximate self regulating dynamic process behaviour has the form:

$$Tp\frac{dPV(t)}{dt} + PV(t) = Kp \cdot CO(t - \theta p)$$

Yet this model cannot describe the kind of integrating process behaviour shown **in these examples.** Such behaviour is better described with the *FOPDT Integrating model* form:

$$\frac{dPV(t)}{dt} = Kp * \cdot CO(t - \theta p)$$

It is interesting to note when comparing the two models above that the FOPDT Integrating form does not have the lone "+ PV" term found on the left hand side of the FOPDT dynamic model.

Also, individual values for the familiar process gain, Kp, and process time constant, Tp, are not separately identified for the FOPDT Integrating model. Instead, an integrator gain, Kp*, is defined that has units of the ratio of the process gain to the process time constant, or:

$$Kp* [=] \frac{Kp}{Tp} \quad \text{or} \quad Kp* [=] PV/(CO.time)$$

Tuning Correlations for Integrating Processes

Analogous to the FOPDT investigations on this site (*e.g.*, here and here), we will see that the FOPDT Integrating model parameters Kp* and Θp of Step 3 can be computed using a **graphical analysis of plot data** or by automated analysis using commercial software.

Step 4 then provides tuning values for controllers such as the dependent, ideal PI form:

$$CO = CO_{bias} + Kc \cdot e(t) + \frac{Kc}{Ti} \int e(t)dt$$

and the dependent, ideal PID form:

$$CO = CO_{bias} + Kc \cdot e(t) + \frac{Kc}{Ti} \int e(t)dt - Kc \cdot Td \frac{dPV}{dt}$$

One important difference about integrating processes is that since there is no identifiable process time constant in the FOPDT Integrating model, we use dead time, Θp, as the baseline marker of time in the design and tuning rules.

Specifically, Θp is used as the basis for computing sample time, T, and the closed loop time constant, Tc. Following the procedures widely for self regulating processes (*e.g.* here and here), we employ a rule to compute the closed time constant, Tc, as:

   $Tc = 3\Theta p$

The controller tuning correlations for integrating processes use this Tc, as well as the Kp* and Θp from the FOPDT integrating model fit, as:

	Controller Gain Kc	Reset Time Ti	Deriv Time Td
PI	$\dfrac{1}{Kp^*} \dfrac{2Tc + \Theta p}{(Tc + \Theta p)^2}$	$2Tc + \Theta p$	
PID	$\dfrac{1}{Kp^*} \dfrac{2Tc + \Theta p}{(Tc + 0.5\Theta p)^2}$	$2Tc + \Theta p$	$\dfrac{0.25\Theta p^2 + Tc\Theta p}{2Tc + \Theta p}$

Loop Sample Time, T

Determining a proper sample time, T, for integrating processes is somewhat more challenging than for self regulating processes.

There are two sample times, T, used in process controller design and tuning. One is the control loop sample time that specifies how often the controller samples the measured process variable (PV) and computes and transmits a new controller output (CO) signal. The other is the rate at which CO and PV data are sampled and recorded during a bump test (step 2 of the recipe).

All controllers measure, act, then wait until next sample time before repeating the loop. This "measure, act, wait" procedure has a delay (or dead time) of one sample time built naturally into its structure. Thus, the minimum dead time (Θp, min) in any control loop is the loop sample time, T.

With this information, we recognize a somewhat circular argument in defining sample time for integrating processes:

our time basis for controller design is Θp, and as such, then loop sample time, T, should be small relative to dead time, or:

$$T \le 0.1\Theta p$$

but the minimum that dead time can be is one sample time, T, or

$$\Theta p, min = T$$

Thus, T is based on Θp, and if the process is sampled too slowly during a bump test, then Θp can be based on T.

To avoid this issue, it is best practice to sample the process as fast as reasonably possible during bump tests so accurate model parameters can be determined during analysis. Loop sample time, T, can then be computed from dead time for controller implementation.

If there is concern about a particular analysis, an alternative and generally conservative way to compute sample time is:

$$T = 0.1 \frac{(PV_{max} - PV_{min})}{(CO_{max} - CO_{min})} \cdot \frac{1}{\left| K_p^* \right|}$$

where the subscripts max and min refer to the maximum and minimum values for CO and PV across the signal span of the instrumentation.

Using the Recipe

The tuning recipe for integrating processes has important differences from that used for self regulating process. When designing and tuning controllers for such processes, we should:

- use an FOPDT Integrating model form when approximating dynamic model behaviour,
- note that the closed loop time constant, Tc, and sample time, T, are based on model dead time, Θp.
- employ PI and PID tuning correlations specific to integrating processes.

WHEN TO USE WHICH TUNING RULE

There are more than 400 tuning rules for PI and PID controllers. How can one possibly choose the best or most appropriate tuning rule from all of these? To simplify matters, the main differences between the tuning rules can be grouped into four categories:

1. Type of process
2. Tuning objective
3. Process information required
4. Type of controller

Most of the tuning rules apply to first-order plus dead time (self-regulating) and integrator plus dead time (integrating) process types. These two process types adequately cover the vast majority of control loops in process plants. Other tuning rules apply to higher-order, oscillating, or unstable processes. Most of the documented tuning rules apply only to processes with dominant time constants. This limits their practical application. The Cohen-Coon tuning rules are an exception.

Tuning objectives include quarter-amplitude damping, minimization of some error integral, a specific percentage overshoot, critically damped, robust tuning, and a specified closed-loop time constant. It is rare to find a tuning rule with an adjustable tuning factor that allows you to change the speed of response. The IMC / Lambda tuning rules are one exception.

The process information required for the tuning rules based on first-order plus dead time and integrator plus dead time process types can be obtained by doing process step tests. A few tuning rules are based on the ultimate cycling or relay tuning methods. Many of the academic tuning rules are based on high-order process models, but they never tell you how to obtain the process model; they just base the tuning on some fictitious model chosen by the author, which largely makes them useless for practical application.

Most tuning-rule authors developed tuning rules for both PI and PID controllers, but with no guidance when to use which one. Some PID tuning rules apply to the interactive algorithm, while most apply to the noninteractive algorithm. It is reasonably easy to convert from one type to the other.

To reduce all these complexities to something we can work with on most control loops, we can consider two process types (self-regulating and integrating), and two tuning objectives (fast and slow or very robust). And ideally we need an easy tuning factor to adjust the speed of response.

When to Use Which Tuning Rule

You could probably use any of the 400 tuning rules, as long as it applies to your situation. I have successfully tuned most (but not all) control loops using just a few tuning rules. Here is what I recommend for most loops:

For self-regulating processes, use the Cohen-Coon PI tuning rule with the following exceptions:

- Use a stability margin of two or more to improve robustness and adjust speed of response.
- If td > 4tau, use the tuning rule for dead-time-dominant processes.
- If you find it difficult to accurately measure the dead time, use the Lambda tuning rule.
- If you want the loop to have a specific speed of response, use the Lambda tuning rule.
- If you want the loop to absorb disturbances rather than pass them on to the next process, use the Lambda tuning rule with the closed loop time constant set three tomes the open loop time constant.
- Use the derivative control mode (PID tuning rule) only when you need every last bit of speed, and then only when the process lends itself well to the use of derivative.

For integrating processes, use the Ziegler-Nichols tuning rule, except for surge tanks and level averaging, where you should use the two tuning rules named after these control objectives.

	Fast Response	Slow / Robust Response
Self-Regulating Process	Cohen-Coon (adjust the stability margin (SM) to change the speed of response)	Lambda (adjust the closed-loop time constant to change the speed of response)
Integrating Process	Ziegler-Nichols (adjust the stability margin (SM) to change the speed of response)	Level-averaging (adjust the specification for maximum deviation from setpoint)

If you use a PID tuning rule and an interactive controller algorithm, or a controller with the parallel algorithm, remember to convert the calculated tuning parameters to ones suitable for your controller algorithm. Also remember to measure your process characteristics in the same time-units your controller's integral uses. And remember to integral time to integral gain – if that is what your controller uses. Finally, when tuning any control loop, watch out for control valve problems.

LEVEL CONTROLLER TUNING

Level control loops are common in industrial processes, but tuning level controllers can be challenging. Many level loops oscillate, sometimes causing large parts of their adjacent processes to oscillate with them

An important thing we need to know about level loops is that liquid level in a vessel is an integrating process, which responds differently from a self-regulating process. Therefore it has a different process model that requires a different set of tuning rules.

Level controller tuning really is not all that difficult if you follow a few basic steps. There are always a few outliers, but in general I like tuning level loops and find them reasonably easy to tune. If the level controller output cascades to a flow controller (more info here), you have to tune the flow control loop first. I'll assume you have done that already and are now ready to tune the level loop.

You should tune any controller based on the process' dynamic response. Obtaining a model for the dynamic response of a tank's level is easy:

- Make sure as far as possible that the uncontrolled flow into/out of the vessel is as constant as possible.
- Place the level controller in manual control mode.
- Wait for a steady slope in the level. If the level is volatile, wait long enough to be able to confidently draw a straight line though the general slope of the level.
- Make a step change in the controller output. Try to make the step change 5% to 10% in size, if the process can tolerate it.

- Wait for the level to change its slope and settle into a new direction. If the level is volatile, wait long enough to be able to confidently draw a straight line though the general slope of the level.
- Restore the level to an acceptable operating point and place the controller back in auto.

Now determine the process model:

- Draw a line (Slope 1) through the initial slope, and extend it to the right.
- Draw a line (Slope 2) through the final slope, and extend it to the left to intersect Slope 1.
- Measure the time between the beginning of the change in controller output and the intersection between Slope 1 and Slope 2. This is the process dead time (td), the first parameter you require for tuning the controller.
- Note: Express your dead time measurement in the same time-base your controller uses for its integral time setting, *i.e.* minutes or seconds.
- Pick any two points (PV1 and PV2) on Slope 1, located conveniently far from each other to make accurate measurements.
- Pick any two points (PV3 and PV4) on Slope 2, located conveniently far from each other to make accurate measurements.
- Calculate the difference in the two slopes as follows:
- $DS = (PV4 - PV3)/T2 - (PV2 - PV1)/T1$
- Note: Express your T1 and T2 measurements in the same time-base your controller uses for its integral time setting, *i.e.* minutes or seconds.
- If your PV is not ranged 0 – 100 %, convert DS to a percentage of the range as follows:
- $DS\% = 100 \times DS / (PV \text{ range max} - PV \text{ range min})$
- Calculate the process integration rate (ri) which is the second and final parameter you need for tuning the controller:
- $ri = DS\% / dCO$

Now that you have the dead time (td) and the process integration rate (ri), you can tune the controller. If the control objective is a nice and fast response to quickly recover from disturbances, you can use a modification of the Ziegler-Nichols (Z/N) tuning rules. The modification involves a slight detuning of the controller because the original Z/N tuning rules result in a very aggressive loop response and low tolerance for any change in operating conditions. I call the amount of detuning the stability margin, denoted by SM. You should set SM to a value of 2.0 or larger. The larger you make SM, the slower the loop will respond. In this way you can use SM as a fine-tuning factor.

Note:

- The tuning rules below assume your controller's proportional setting is in gain Kc, not Proportional Band, PB. If not: $PB = 100 / Kc$.

- The tuning rules below also assume your controller's integral setting is in units of time Ti (i.e minutes or seconds), not repeats per time Ki. If not: Ki = 1 / Ti.
- The tuning rules below also assume you have a controller with an interacting algorithm (although they work fairly well on noninteracting algorithms too), but not a parallel algorithm. For controllers with the parallel algorithm, you need to divide Ti by Kc, and multiply Td by Kc, to obtain their integral and derivative settings, respectively.

To calculate tuning constants for a PI controller:

Kc = 0.9 / (SM x ri x td)

Ti = 3.33 x SM x td

Td = 0

And for a PID controller:

Kc = 1.2 / (SM x ri x td)

Ti = 2 x SM x td

Td = td / 2

Important Note:

Some level controllers should not respond fast, *e.g.* when controlling the level of a surge tank. Surge tanks need a different set of tuning rules to ensure you make maximum use of the surge capacity, while not exceeding the upper and lower level limits.

PID TUNING RULES

Safety, equipment and environmental protection, process efficiency and capacity, product quality, and control system maintenance depend on PID tuning.

Nearly every automation system supplier, consultant, control theory professor, and user has a favorite set of PID tuning rules. Many of these experts are convinced their set is the best. A handbook devoted to tuning has over 500 pages of rules. The enthusiasm and sheer number of rules is a testament to the importance of tuning and the wide variety of application dynamics, requirements, and complications. The good news is these methods converge for a common objective. The addition of PID features, such as setpoint lead-lag, dynamic reset and output velocity limits, and intelligent suspension of integral action enable the use of disturbance rejection tuning to achieve other system requirements, such as maximizing setpoint response, coordinating loops, extending valve packing life, and minimizing upsets to operations and other control loops.

Potential Performance

The purpose of a control loop is to reject undesired changes, ignore extraneous changes, and achieve desired changes, such as new setpoints. PID control provides the best possible rejection of unmeasured disturbances (regulatory control) when properly tuned. The addition of a simple deadtime block in the external reset path can enhance the PID regulatory control capability more than other controllers with intelligence built-in to process dynamics, such as model predictive control. In plants, unknown and extraneous changes are a reality, and the PID is the best tool if properly tuned. The test time has been significantly reduced for the most difficult loops. Simple equations have been developed to estimate tuning and resulting performance for a unified approach. (Equation derivations and a simple tuning method are in the online version.)

Control Requirements

The foremost requirement of a PID is to prevent the activation of a safety instrumentation system or a relief device and the prevention of an environmental violation (RCRA pH), compressor surge, and shutdown from a process excursion. The peak error (maximum deviation from setpoint) is the most applicable metric. The most disruptive upset is an unmeasured step disturbance that would cause an open loop error (E_o) if the PID was in manual or did not exist. The fraction of open loop error seen in feedback control is more dependent upon the controller gain than the integral time since the proportional mode provides the initial reaction important for minimizing the peak error. Equation shows if the product of the controller gain (K_c) and open loop gain (K_o) is much greater than one, the peak error (E_x) is significantly less than the open loop error. The open loop gain (K_o) is the product of the final element, process, and measurement gain and is the percent change in process variable divided by the percent change in controller output for a setpoint change. For most vessel and column temperature and pressure control loops, the process rate of change is much slower than the deadtime. Consequently, the controller gain can be set large enough where the denominator becomes simply the inverse of the product of the gains. Conversely, for loops dominated by deadtime, the denominator approaches one, and the peak error is essentially the open loop error.

$$E_x = \frac{1}{(1 + K_c * K_o)} * E_o$$

The peak error is critical for product quality in the final processing of melts, solids, or paste, such as extruders, sheet lines, and spin lines. Peak errors show up as rejected product due to colour, consistency, optical clarity, thickness, size, shape, and in the case of food, palatability. Unfortunately, these systems are dominated by transportation delays. The peak errors and disruptions from upstream processes must be minimized.

The most widely cited metric is an integrated absolute error (IAE), which is the area between process variable and the setpoint. For a non-oscillatory response, the IAE and the integrated error (IE) are the same. Since proportional and integral action are important for minimizing this error, Equation shows the IE increases as the integral time (T_i) increases and the controller gain decreases.

$$E_i = \left[\frac{(T_i + \Delta t_x + t_f)}{K_o * K_c} \right] * E_o$$

Equation also shows how the IE increases with controller execution time (Δt_x) and signal filter time (τ_f). The equivalent deadtime from these terms also decreases the minimum allowable integral time and maximum allowable controller gain, further degrading the maximum possible performance. In many cases, the original controller tuning is slower than allowed and remains unchanged, so the only deterioration observed is from these terms in the numerator of Equation. Studies on the effect of automation system dynamics and innovations can lead to conflicting results because of the lack of recognition of the effect of tuning on the starting case and comparative case performance. In other words, you can readily prove anything you want by how you tune the controller.

IE is indicative of the quantity of product that is off-spec that can lead to a reduced yield and higher cost ratio of raw material or recycle processing to product. If the off-spec cannot be recycled or the feed rate cannot be increased, there is a loss in production rate. If the off-spec is not recoverable, there is a waste treatment cost.

A controller tuned for maximum performance will have a closed loop response to an unmeasured disturbance that resembles two right triangles placed back to back. The base of each triangle is the total loop deadtime and the altitude is the peak error. If the integral time (reset time) is too slow, there is slower return to setpoint. If the controller gain is too small, the peak error is increased, and the right triangle is larger for the return to setpoint.

Process Dynamics

The major types of process dynamics are differentiated by the final path of the open loop response to a change in manual controller output assuming no disturbances. (The online version shows the three major types of responses and

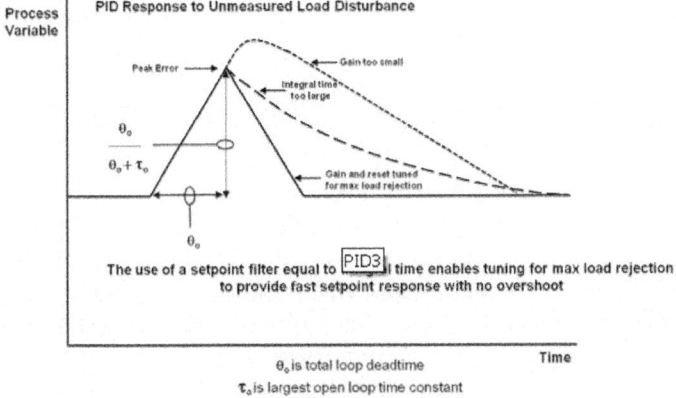

Fig. : The effect of integral time on the maximum possible disturbance rejection.

the associated dynamic terms.) If the response lines out to a new steady state, the process is self-regulating with an open loop time constant (τ_o) that is the largest time constant in the loop. Flow and continuous operation temperature and concentration are self-regulating processes. If the response continues to ramp, the process is integrating. Level, column and vessel pressure, batch operation temperature, and concentration are integrating processes. If the response accelerates, reaching a point of no return, the process has positive feedback leading to a runaway. Batch or continuous temperature in highly exothermic reactors (*e.g.*, polymerization) can become runaway processes. Prolonged open loop tests are not permitted, and setpoint changes are limited. Consequently, the acceleration is rarely intentionally observed.

Unified Approach

The three major types of responses have an initial period of no response that is the total loop deadtime (θ_o) followed by the ramp before the deceleration (inflection point) of a self-regulating response and the acceleration of the runaway response. The percent ramp rate divided by the change in percent controller output is the integrating process gain (K_i) with units of %/sec/%, which reduces to 1/sec.

For at least 10 years, slow self-regulating processes with a long time to deceleration have shown to be effectively identified and tuned as "near integrating" or "pseudo integrating" processes, leading to a "short cut tuning method" where only the deadtime and initial ramp rate need to be recognized. The tuning test time for these "near integrating" processes can be reduced by over 90% by not waiting for a steady state. Recently, the method was extended to runaway processes and to deadtime dominant self-regulating processes by the use of a deadtime block to compute the ramp rate over a deadtime interval. Furthermore, other tuning rules were found to give the same equation for controller gain when the performance objective was maximum unmeasured disturbance rejection. For example, the use of a closed loop time constant (λ) equal to the total loop deadtime in Lambda tuning yields the same result as the Ziegler Nichols (ZN) ultimate oscillation and reaction curve methods if the ZN gain is cut in half for smoothness and robustness. Equation shows the controller gain is half the inverse of the product of integrating process gain and deadtime.

$$K_c = \frac{0.5}{K_i * \theta_o}$$

The profession realizes that too large of a controller gain will cause relatively rapid oscillations and can instigate instability (growing oscillations). Unrealized for integrating process is that too small of a controller gain can cause extremely slow oscillations that take longer to decay as the gain is decreased. Also unrealized for a runaway process is that a controller gain set less than the inverse of the open loop gain causes an increase in temperature to accelerate to a point of no return. There is a window of allowable controller gains. Also realized is too small of an integral time will cause overshoot and can lead to a reset cycle.

Almost completely unrealized is that too slow of an integral time will result in a sustained overshoot of a setpoint that gets larger and more persistent as the integral time is increased for integrating processes. Hence a window of allowable integral times exists. Equation provides the right size of integral time for integrating processes. If we substitute Equation into Equation, we end up with Equation, which is a common expression for the integral time for maximum disturbance rejection. Equation is extremely important because most integrating processes have a controller gain five to 10 times smaller than allowed. The coefficient in Equation can be decreased for self-regulating processes as the deadtime becomes larger than the open loop time constant (τ_o) estimated by Equation.

$$T_i = \frac{2}{(K_c * K_i)}$$

$$T_i = 4 * \theta_o$$

$$\tau_o = \frac{K_o}{K_i}$$

The tuning used for maximum load rejection can be used for an effective and smooth setpoint response if the setpoint change is passed through a lead-lag. The lag time is set equal to the integral time, and the lead time is set approximately equal to ¼ the lag time.

For startup, grade transitions, and optimization of continuous processes and batch operations, setpoint response is important. Minimizing the time to reach a new setpoint (rise time) can in many cases maximize process efficiency and capacity. The rise time (T_r) for no output saturation, no setpoint feedforward, and no special logic is the inverse of the product of the integrating process gain and the controller gain plus the total loop deadtime. Equation is independent of the setpoint change.

$$T_r = \frac{1}{(K_i * K_c)} + \theta_o$$

Complications, Easy Solutions

Fast changes in controller output can cause oscillations from a slow secondary loop or a slow final control element. The problem is insidious in that oscillations may only develop for large disturbances or large setpoint changes. The enabling of the dynamic reset limit option and the timely external reset feedback of the secondary loop or final control element process variable will prevent the primary PID controller output from changing faster than the secondary or final control element can respond, preventing oscillations.

Aggressive controller tuning can also upset operations, disturb other loops, and cause continual crossing of the split range point. Velocity limits can be added to the analog output block, the dynamic reset limit option enabled, and the block process variable used as the external reset to provide directional move suppression to smooth out the response as necessary without retuning.

The different closed loop response of loops can reduce the coordination, especially important for blending and simplification of the identification of models for advanced process control systems that manipulate these loops. Process nonlinearities may cause the response in one direction to be faster. Directional output velocity limits and the dynamic reset limit option can be used to equalize closed loop time constants without retuning.

Final control element resolution limits (stick-slip) and deadband (backlash) can cause a limit cycle if one or two or more integrators, respectively, exist in the loop. The integrator can be in the process or in the secondary or primary PID controller via the integral mode. Increasing the integral time will make the cycle period slower but cannot eliminate the oscillation. However, a total suspension of integral action when there is no significant change in the process variable and when the process is close to the setpoint can stop the limit cycle. The output velocity limits can also be used to prevent oscillations in the controller output from measurement noise exceeding the deadband or resolution limit of a control valve preventing dither, which further reduces valve wear.

Bottom Line

Controllers can be tuned for maximum disturbance rejection by a unified method for the major types of processes. PID options in today's DCS, such as setpoint lead-lag, directional output velocity limits, dynamic reset limit, and intelligent suspension of integral action, can eliminate oscillations without retuning. Less oscillations reduces process variability, enables better recognition of trends, offers easier identification of dynamics, and provides an increase in valve packing life.

TANK LEVEL TUNING COMPLICATIONS

Level control loops are strange creatures. This strangeness can make them difficult to tune. On average, level control loops are tuned the worst of all process types. Although I have seen poorly tuned loops of all types, poorly tuned level controllers typically have tuning settings that are the furthest from optimal. Most level processes are very robust in nature, allowing them to function surprisingly well with suboptimal tuning.

But it does not have to be this way. If controller tuning is based on the dynamic response of a process, most level control loops are actually easy to tune and provide very robust control. However, as you probably know, most control loops are tuned "intuitively" using trial and error. More often than not, this approach results in poor control loop performance.

Case Study

A few weeks ago, I helped an engineer at a power plant with the tuning of a demineralized (demin) water storage tank. It was a large tank – about 40 feet

(12 m) high and 20 feet (6 m) in diameter. Water was pumped from the demin water production plant into the tank, and this flow rate was manipulated with a control valve. Under normal operating conditions the unit consumesd demin water at an almost constant rate (most of which was discharged through the continuous boiler blowdown).

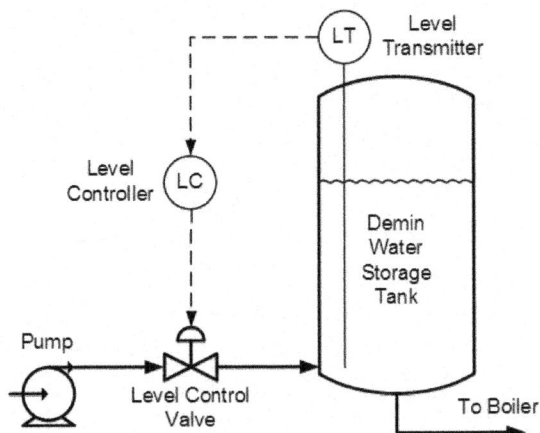

Fig. : Demineralized water storage tank level control.

To do the tuning correctly, the engineer executed a few step tests and we analyzed the data. We calculated the process integration rate (or process gain) to be 0.0045 / minute. This means if the level is at steady state and the controller output is changed manually by X percent, it will take 1/0.0045 minutes (3.7 hours) for the level to change by the same percentage. The dead time was measured to be roughly 2.5 minutes.

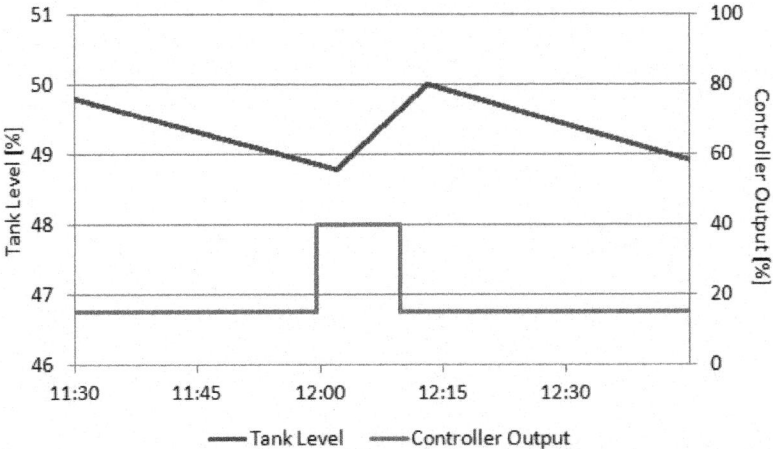

Fig. : The two step tests used for tuning.

Once we had this information on the dynamic properties of the process, we used the modified Ziegler-Nichols tuning rules for Integrating Processes and calculated new tuning settings for this control loop. We used a "stability margin" of 2.5 and obtained the following tuning settings:

Controller Gain (Kc) = 32

Integral Time (Ti) = 20 minutes.

The high controller gain was a concern. Although the level was quite smooth during our step tests, a historical trend of level revealed some jittering was present at times. And since a 1% jitter in level would cause the controller output to "jitter" by 32% (Kc x delta PV), we decided to use a lower controller gain since tight control was not a requirement. We felt that Kc = 10 would be a good compromise between control performance and jitter tolerance.

Tuning Complications

Many level loops have small integration rates (or process gains). Integration rate (ri) is inversely proportional to the vessel's residence time. Typically, the larger the tank, the smaller the integration rate. The process with the smallest integration rate that I personally worked with was a city water reservoir, which had a residence time of 48 hours (ri = 0.000347 / minute). For good control, a very low integration rate theoretically requires a very high controller gain, sometimes in excess of 100. Practically we cannot use controller gains of this magnitude because of the severe control action that would result from noise and setpoint changes. (Note that one can also overcome severe control action by using a noise filter and either the P&D-on-error control algorithm or a setpoint filter).

This mandatory reduction of the controller gain brings me to the reason why most level loops have grossly suboptimal tuning settings. For integrating control loops (such as tank level), when you reduce the controller gain you have to increase the integral time, otherwise the loop can become very oscillatory.

Unenlightened tuners do not know of this requirement and end up using disproportionately short integral times on level loops, resulting in very oscillatory behaviour. When they try to stabilize the loop by further reducing the controller gain, the situation deteriorates even more.

Example

For example, let's look at how Billy, our unenlightened but fictitious tuner, might have tuned the tank level controller. Assuming he did step tests, he then used the original Ziegler-Nichols tuning rules (I did mention he is unenlightened) for calculating the controller settings. He obtained the following controller settings: Controller Gain (Kc) = 80 and Integral Time (Ti) = 8.3 minutes. He realized that the controller gain of 80 was too high, and reduced it to 10. But he left the integral time at 8.3 minutes, as calculated.

Then he tested the new tuning settings and noticed overshoot and oscillations in level. Too much gain, right? So he set the Kc value to 5 and retested the performance. The loop still oscillated with the adjusted tuning settings, but he realized that this tuning effort was taking too much of his time, so he left the tuning settings as they were and moved on to other work.

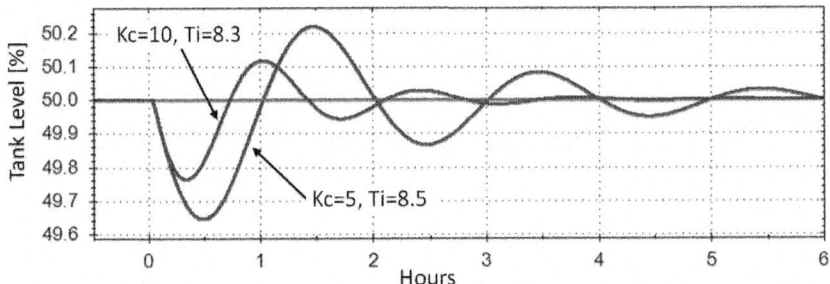

Fig. : The result of using decreased controller gains on a level loop, while leaving the integral time at the originally calculated value.

How it's Done

Now back to our own tuning efforts on the demin water storage tank. When the engineer and I reduced the controller gain from 32 to 10, we simultaneously increased the integral time from 20 to 64 minutes, which we calculated using the equation below.

Equation for calculating a new integral time when reducing the controller gain in a level loop:

Ti(new) = Ti(old) x Kc(old) / Kc(new)

The level loop's response to a 5% change in outflow using the initial and refined controller settings. The control loop is significantly more stable compared to the alternatives.

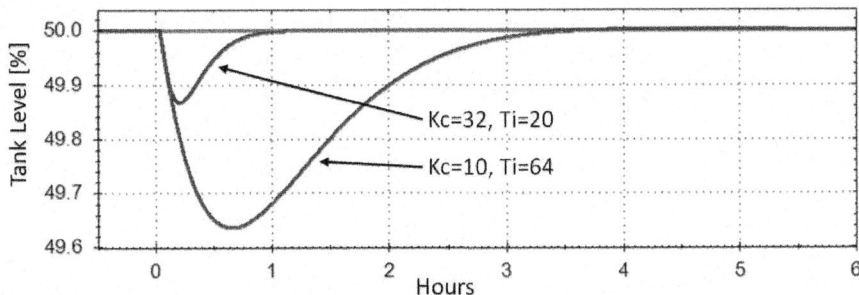

Fig. : Stable level control loop response obtained from increasing integral time while decreasing controller gain.

As I said at the beginning, level controller tuning does not have to be difficult. Do step-tests to understand the process dynamics, use proven tuning rules to calculate controller settings, and remember to adjust the integral time inversely to any subsequent change you make in controller gain.

SURGE TANK LEVEL CONTROL

A surge tank is placed between two processing units to absorb flow rate fluctuations coming from the upstream process and keep the flow rate to the downstream process more constant. To do this, the tank level has to go up and down. Consequently, the level controller should not try to hold the level as close as possible to its set point; the controller should simply keep the surge tank's level between its upper and lower limits, and do this with the least possible amount of change to its output.

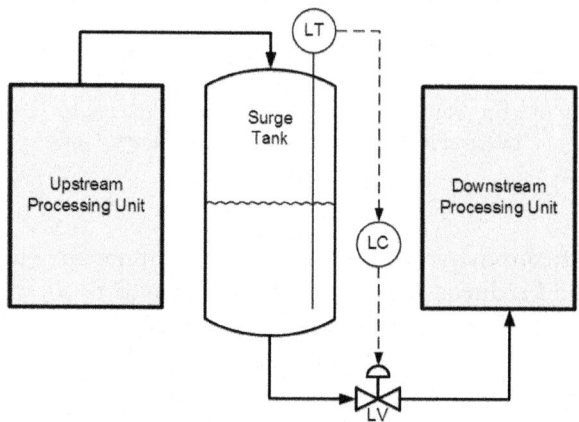

Fig. : Surge tank level control loop.

Although there are other methods of controlling surge tank level, the level-averaging method is preferred by most operators and process engineers. This method minimizes control valve movement during disturbances, keeps the level between its limits, and brings the level back to setpoint in the long term. Another method of surge tank control does not bring the level back to setpoint but potentially provides more surge capacity; I'll write about that method another day.

Surge Tank Level Controller Tuning

To tune the controller for level-averaging control, you need to know the following three things:

1. The residence time of the vessel (t_{res})

The residence time is the time it would take for the surge tank to drain from 100% level to 0% level if there is no flow into the tank and the outlet valve is 100% open. You can calculate this as the volume of liquid contained in the vessel between 0% and 100% of the span of its level measurement, divided by the maximum flow rate with the outlet valve wide open: $t_{res} = V/Q_{max}$. Use the same engineering unit for volume in V and Q_{max}. If you don't know the volume and/or maximum flow rate, you can estimate the residence time as the inverse of the process integration rate, $t_{res} = 1/r_i$. You can determine r_i through step testing. Be sure to express t_{res} in the same time-base as your controller's integral time (minutes versus seconds).

2. The largest expected change in flow rate (Δf_{max}).

This should be expressed as a percentage of maximum valve capacity. You can review historical trends of the loop and find the largest change the controller output has made (under automatic control) to control the level.

3. The maximum tolerable deviation from setpoint (ΔL_{max}).

This should be expressed as a percentage of the span of the level measurement.

Once you have all of these, calculate tuning settings for the controller with the equations below.

For a controller with an interactive or noninteractive algorithm:

$K_C = 0.74 \, \Delta f_{max} / \Delta L_{max}$

$T_I = 4 \, t_{res} / K_C$

$T_D = 0$

K_C is controller gain.

If your controller uses proportional band, $PB = 100/K_C$.

T_I is integral time in the same units as t_{res}.

If your interactive or noninteractive controller uses integral gain, $K_I = 1/T_I$.

T_D is the derivative time.

For a controller with a parallel algorithm:

$K_P = 0.74 \, \Delta f_{max} / \Delta L_{max}$

$K_I = K_P^2 / (4 \, t_{res})$

$K_D = 0$

K_P is proportional gain and K_I is integral gain using the same time-base as t_{res}.

If your parallel controller uses integral time: $T_I = 1 / K_I$

K_D is the derivative time.

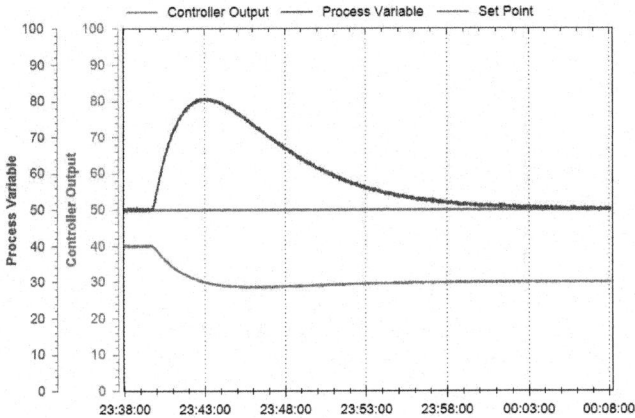

Fig. : Response of a surge tank level control loop to a disturbance in inlet flow rate.

Faster tuning is also possible. The following equations will produce tuning settings to bring the level back towards the setpoint much quicker. The level will slightly overshoot the setpoint as a result of the faster response.

For a controller with an interactive or noninteractive algorithm:

$$K_C = 0.5\ \Delta f_{max}\ /\ \Delta L_{max}$$
$$T_I = 0.74\ t_{res}\ /\ K_C$$
$$T_D = 0$$

For a controller with a parallel algorithm:

$$K_P = 0.5\ \Delta f_{max}\ /\ \Delta L_{max}$$
$$K_I = K_P^2\ /\ (0.74\ t_{res})$$
$$K_D = 0$$

The same parameter descriptions and conversions given previously, apply to the faster tuning equations too.

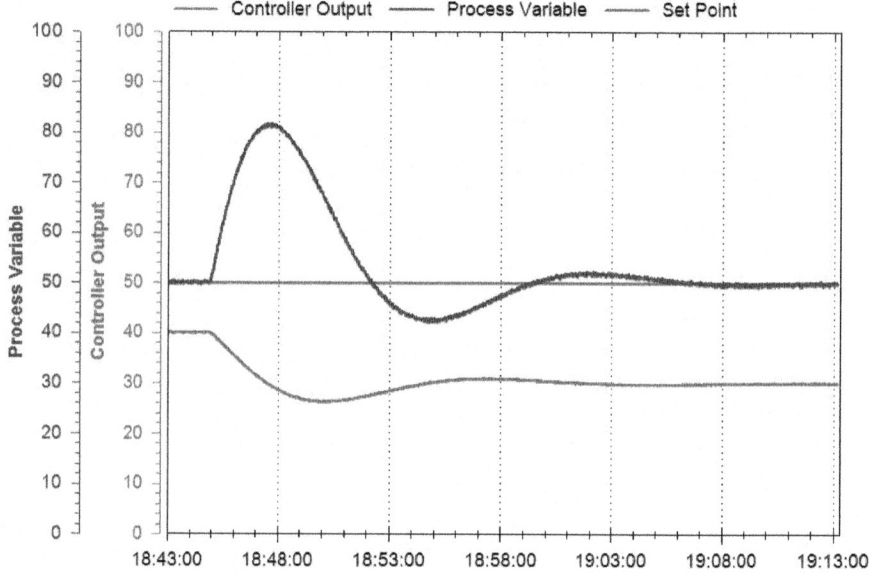

Fig. : A faster response to the same disturbance in inflow.

With these tuning rules, you should be able to get your surge tanks under control, and have them respond appropriately to surges in inflow. Let me know if you need help.

DESIGN AND TUNING RECIPE MUST CONSIDER NONLINEAR PROCESS BEHAVIOUR

Processes with streams comprised of gases, liquids, powders, slurries and melts tend to exhibit variations in behaviour as operating level changes. This, in fact, is the very nature of a nonlinear process. For this reason, our recipe for controller design and tuning begins by specifying our design level of operation.

Controller Design and Tuning Recipe:

1. Establish the design level of operation (DLO), which is the normal or expected values for set point and major disturbances.

2. Bump the process and collect controller output (CO) to process variable (PV) dynamic process data around this design level.

3. Approximate the process data behaviour with a first order plus dead time (FOPDT) dynamic model.

4. Use the model parameters from step 3 in rules and correlations to complete the controller design and tuning.

Nonlinear Behaviour of the Gravity Drained Tanks

The dynamic behaviour of the gravity drained tanks process is reasonably intuitive. Increase or decrease the inlet flow rate into the upper tank and the liquid level in the lower tank rises or falls in response.

One challenge this process presents is that its dynamic behaviour is nonlinear. That is, the process gain, Kp; time constant, T_p; and/or dead time, Θ_p; changes as operating level changes. This is evident in the open loop response plot below.

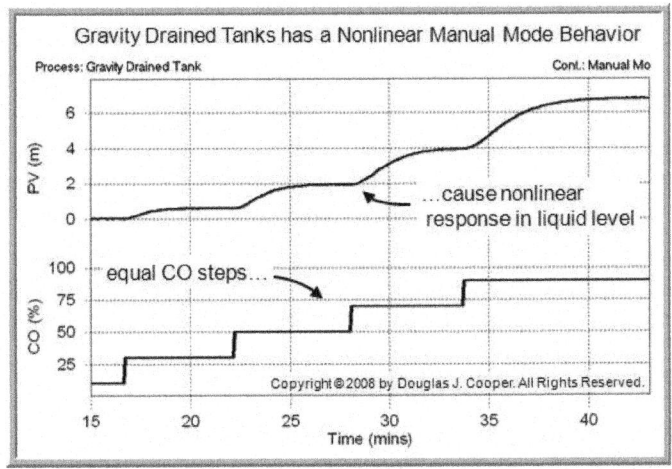

As shown above, the CO is stepped in equal increments, yet the response behaviour of the PV changes as the level in the tank rises. The consequence of nonlinear behaviour is that a controller designed to give desirable performance at one operating level may not give desirable performance at another level.

Nonlinear Behaviour of the Heat Exchanger

Nonlinear process behaviour has important implications for controller design and tuning. Consider, for example, our heat exchanger process under PI control.

When tuned for a moderate response as shown in the first set point step from 140 °C to 155 °C in the plot below, the process variable (PV) responds in a manner consistent with our design goals. That is, the PV moves to the new set point (SP) reasonably quickly but does not overshoot the set point.

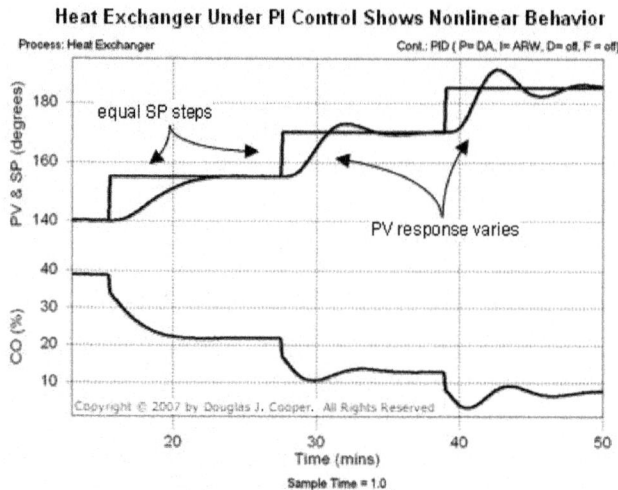

Heat Exchanger Under PI Control Shows Nonlinear Behavior

The consequence of a nonlinear process character is apparent as the set point steps continue to higher temperatures. In the third set point step from 170 °C to 185 °C, the same controller that had given a desired moderate performance now produces a PV response with a clear overshoot and some oscillation.

Such a change in performance with operating level may be tolerable in some applications and unacceptable in others. The "best" performance is something we judge for ourselves based on the goals of production, capabilities of the process, impact on down stream units and the desires of management

Nonlinear behaviour should not catch us by surprise. It is something we can know about our process in advance. And this is why we should choose a design level of operation as a first step in our controller design and tuning procedure.

Step 1: Establish the Design Level of Operation (DLO)

Because, as shown in the examples above, processes have process gain, Kp; time constant,Tp; and/or dead time, Θp values that change as operating level changes, and these FOPDT model parameter values are used to complete the controller design and tuning procedure, it is important that dynamic process test data be collected at a pre-determined level of operation.

Defining this design level of operation (DLO) includes specifying where we expect the set point (SP) and measured process variable (PV) to be during normal operation, and the range of values the SP and PV might typically assume. This way we know where to explore the dynamic process behaviour during controller design and tuning.

The DLO also considers our major disturbances (D). We should know the normal or typical values for our major disturbances. And we should be reasonably confident that thedisturbances are quiet so we may proceed with a bump test to generate and record dynamic process data.

Step 2. Collect Dynamic Process Data Around the DLO

The next step in our recipe is to collect dynamic process data as near as practical to our design level of operation. We do this with a bump test, where we step or pulse the CO and collect data as the PV responds.

It is important to wait until the CO, PV and D have settled out and are as near to constant values as is possible for our particular operation before we start a bump test. The point of bumping a process is to learn about the cause and effect relationship between the CO and PV.

With the process at steady state, we are starting with a clean slate. As the PV responds to the CO bumps, the dynamic cause and effect behaviour is isolated and evident in the data. On a practical note, be sure the data capture routine is enabled before the initial bump is implemented so all relevant data is collected.

Two popular open loop (manual mode) methods are the step test and the doublet test.

For either method, the CO must be moved far enough and fast enough to force a response in the PV that dominates the measurement noise.

Also, our bump should move the PV both above and below the DLO during testing. With data from each side of the DLO, the model (step 3) will be able to average out the nonlinear effects.

• *Step Test*

To collect data that will "average out" to our design level of operation, we start the test at steady state with the PV on one side of (either above or below) the DLO. Then, as shown in the plot below, we step the CO so that the measured PV moves across to settle on the other side of the DLO.

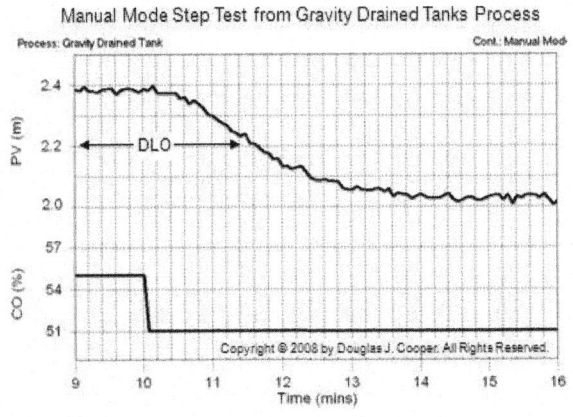

Manual Mode Step Test from Gravity Drained Tanks Process

We can either start high and step the CO down (as shown above), or start low and step the CO up. Both methods produce dynamic data of equal value for our design and tuning recipe.

• *Doublet Test*

A doublet test, as shown below, is two CO pulses performed in rapid succession and in opposite direction. The second pulse is implemented as soon as the process has shown a clear response to the first pulse that dominates the noise in the PV. It is not necessary to wait for the process to respond to steady state for either pulse.

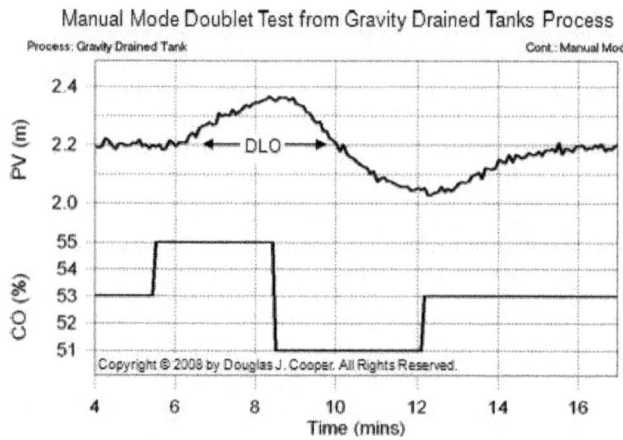

The doublet test offers attractive benefits, including that it starts from and quickly returns to the DLO, it produces data both above and below the design level to "average out" the nonlinear effects, and the PV always stays close to the DLO, thus minimizing off-spec production. Such data does require commercial software for model fitting, however.

Step 3: Fit a FOPDT Dynamic Model to Process Data

In fitting a first order plus dead time (FOPDT) model, we approximate those essential features of the dynamic process behaviour that are fundamental to control. We need not understand differential equations to appreciate, but for completeness, the first order plus dead time (FOPDT) dynamic model has the form:

$$Tp\frac{dPV(t)}{dt} + PV(t) = Kp \cdot CO(t - \theta p)$$

Where:

PV(t) = measured process variable as a function of time

CO(t - Θp) = controller output signal as a function of time and shifted by Θp

Θp = process dead time

t = time

When the FOPDT dynamic model is fit to process data, the results describe how PV will respond to a change in CO via the model parameters. In particular:

- Process gain, Kp, describes the direction and how far PV will travel,
- Time constant, Tp, states how fast PV moves after it begins its response,
- Dead time, Θp, is the delay from when CO changes until when PV begins to respond.

An example study that compares dynamic process data from the heat exchanger with a FOPDT model prediction can be found here. Comparisons between data and model for the gravity drained tanks can be found here and here.

Step 4: Use the Model Parameters to complete the Design and Tuning

In step 4, the three FOPDT model parameters are used in correlations to compute controller tuning values. For example, the chart below lists internal model control (IMC) tuning correlations for the PI controller and dependent ideal PID controller, and dependent ideal PID with CO filter forms:

	Controller Gain Kc	Reset Time Ti	Deriv Time Td	Filter Const α
PI	$\dfrac{1}{Kp}\dfrac{Tp}{(\Theta p + Tc)}$	Tp		
Ideal PID	$\dfrac{1}{Kp}\left(\dfrac{Tp+0.5\Theta p}{Tc+0.5\Theta p}\right)$	$Tp+0.5\Theta p$	$\dfrac{Tp\,\Theta p}{2Tp+\Theta p}$	
PID w/ CO Filter	$\dfrac{1}{Kp}\left(\dfrac{Tp+0.5\Theta p}{Tc+\Theta p}\right)$	$Tp+0.5\Theta p$	$\dfrac{Tp\,\Theta p}{2Tp+\Theta p}$	$\dfrac{Tc(Tp+0.5\Theta p)}{Tp(Tc+\Theta p)}$

The closed loop time constant, Tc, in the IMC correlations is used to specify the desired speed or quickness of our controller in responding to a set point change or rejecting a disturbance. The closed loop time constant is computed:

aggressive performance: Tc is the larger of $0.1 \cdot Tp$ or $0.8\,\Theta p$

moderate performance: Tc is the larger of $1 \cdot Tp$ or $8\,\Theta p$

conservative performance: Tc is the larger of $10 \cdot Tp$ or $80\,\Theta p$

Use the Recipe – It is Best Practice

The FOPDT dynamic model of step 3 also provides us the information we need to decide other controller design issues, including:

Controller Action

Before implementing our controller, we must input the proper direction our controller should move to correct for growing errors. Some vendors use the term "reverse acting" and "direct acting." Others use terms like "up-up" and "up-down" (as CO goes up, then PV goes up or down). This specification is determined solely by the sign of the process gain, Kp.

Loop Sample Time, T

Process time constant, Tp, is the clock of a process. The size of Tp indicates the maximum desirable loop sample time. Best practice is to set loop sample time, T, at 10 times per time constant or faster (T $\leq$ 0.1Tp). Faster may provide modestly improved performance. Slower than five times per time constant leads to significantly degraded performance.

Dead Time Problems

As dead time grows greater than the process time constant (when $\Theta p > Tp$), controller performance can benefit from a model based dead time compensator such as the Smith predictor.

Model Based Control

If we choose to employ a Smith predictor, or perhaps a dynamic feed forward element, a multivariable decoupler, or any other model based controller, we need a dynamic model of the process to enter into the control computer. The FOPDT model from step 3 of the recipe is usually appropriate for this task.

Chapter 7

ADVANCED REGULATORY CONTROL

ADVANCED REGULATORY VERSUS MODEL-PREDICTIVE CONTROL

Some currently practicing control engineers may be regarding the features and capabilities of advanced regulatory control (ARC) as inferior to model-predictive control (MPC) — but are they really?

A contributor to a recent LinkedIn APC blog made the following statement about the difference between advanced regulatory control (ARC) and model-predictive control (MPC):

"In a general and very simple way, an advanced regulatory controller acts based on an error, while a predictive controller uses a model in order to predict what is going to happen and acts in consequence to avoid it."

As a long-time proponent of ARC for the solution of many industrial control problems, I would like to correct a common misconception held by probably 99% of current practicing control engineers regarding the features and capabilities of ARC.

During the 1970s, and with the advent of digital control systems and process control computers, those of us involved in the evolution of ARC developed two very powerful ARC techniques, referred to as feed forward control and decoupling control. The reason for developing these techniques was quite simple: feedback control, regardless of how sophisticated, was insufficient by itself for keeping important control variables close to setpoint when disturbances occurred.

Feedforward control was developed initially as the name suggests — to make adjustments to independent variables (MVs) to keep important dependent variables (CVs) close to setpoint when the feed rate to a unit operation changed. How was this done? With models, of course! Open loop response data was first gathered to observe the response of the CVs to changes in the feed rate. This response was typically fit with a dead time and lag model. The next step was to "invert" this model so as to develop the feedforward pattern needed to adjust the MV used to control the CV so as to keep the CV close to setpoint during the feed rate change.

So, for example, on a distillation column where the overhead temperature is being controlled with reflux flow, the challenge is to apply dead time and lead/lag compensation to the feed rate, passing the incremental trajectory changes (with proper gain) to the reflux flow, so that the reflux is adjusted in coordinated fashion, such that the disturbance is cancelled out. If the feed forward compensation is perfect, no feedback correction is required at all!

MPC provides more or less identical "model-predictive" control action for feed forward variables (FFVs). ARC feedforward control action can actually be made superior to MPC feedforward control action. The "proper gain" — that is, the amount the MV must be adjusted to keep the CV on setpoint — is not a constant ratio. For example, the reflux/feed ratio required to keep the top temperature on setpoint may be different today from what it was yesterday, due to feed composition changes, *etc.* ARC can incorporate dynamic, adaptive feedforward gain compensation to account for process changes. MPC cannot.

The other model-based control feature of ARC, decoupling control, was developed for similar reasons, and initially for distillation columns. On many fractionators where high purity products are being produced, the heat and material balances are highly coupled, such that moves made on one end of the column also affect the other end. We recognized this phenomenon early on and realized that simultaneous decoupling moves could be made on both ends of the column so as to minimize heat and material balance upsets. For example, if the reflux needs to be increased to increase the purity of the overhead product, then a similar decoupling move needs to be made to the reboiler heat so as not to disturb the bottom product purity. The decoupling moves are determined from a model developed in a manner similar to the one described earlier.

Sounds to me like these ARC techniques developed over 40 years ago:

"Use a model in order to predict what is going to happen and act in consequence to avoid it."

Which is to refute the misconception that ARC is not model-predictive. It certainly is as implemented today, and has been for the last 40 years. In addition, it can be made multi-variable; that is, feed forward and decoupling control action can be implemented simultaneously for several FFVs and DCVs acting on several MVs to control several CVs. And, finally, in general, ARC is cheaper to implement, is more durable, is more easily understood by operators, and requires less maintenance, when compared to MPC. It should always be considered first as the go-to technology for solution of most common industrial control problems.

ADVANCED PROCESS CONTROL

In control theory Advanced process control (APC) refers to a broad range of techniques and technologies implemented within industrial process control systems. Advanced process controls are usually deployed optionally and in addition to *basic* process controls. Basic process controls are designed and built with the process itself, to facilitate basic operation, control and automation requirements. Advanced process controls are typically added subsequently, often over the course of many years, to address particular performance or economic improvement opportunities in the process.

Process control (basic and advanced) normally implies the process industries, which includes chemicals, petrochemicals, oil and mineral refining, food processing, pharmaceuticals, power generation, *etc.* These industries are characterized by continuous processes and fluid processing, as opposed to discrete parts manufacturing, such as automobile and electronics manufacturing. The term process automation is essentially synonymous with process control.

Process controls (basic as well as advanced) are implemented within the process control system, which usually means a distributed control system (DCS), programmable logic controller (PLC), and/or a supervisory control computer. DCSs and PLCs are typically industrially hardened and fault-tolerant. Supervisory control computers are often not hardened or fault-tolerant, but they bring a higher level of computational capability to the control system, to host valuable, but not critical, advanced control applications. Advanced controls may reside in either the DCS or the supervisory computer, depending on the application. Basic controls reside in the DCS and its subsystems, including PLCs.

Types of Advanced Process Control

Following is a list of the best known types of advanced process control:

- Advanced regulatory control (ARC) refers to several proven advanced control techniques, such as feed forward, override or adaptive gain. ARC is also a catch-all term used to refer to any customized or non-simple technique that does not fall into any other category. ARCs are typically implemented using function blocks or custom programming capabilities at the DCS level. In some cases, ARCs reside at the supervisory control computer level.

- Multivariable Model predictive control (MPC) is a popular technology, usually deployed on a supervisory control computer, that identifies important independent and dependent process variables and the dynamic relationships (models) between them, and uses matrix-math based control and optimization algorithms, to control multiple variables simultaneously. MPC has been a prominent part of APC ever since supervisory computers first brought the necessary computational capabilities to control systems in the 1980s.

- Inferential control: The concept behind inferentials is to calculate a stream property from readily available process measurements, such as temperature and pressure, that otherwise would require either an expensive and complicated online analyzer or periodic laboratory analysis. Inferentials can be utilized in place of actual online analyzers, whether for operator information, cascaded to base-layer process controllers, or multivariable controller CVs.

- Sequential control refers to dis-continuous time and event based automation sequences that occur within continuous processes. These may be implemented as a collection of time and logic function blocks, a custom algorithm, or using a formalized Sequential function chart methodology.

- Compressor control typically includes compressor anti-surge and performance control.

Related Technologies

The following technologies are related to APC and in some contexts can be considered part of APC, but are generally separate technologies having their own.

- Statistical process control (SPC), despite its name, is much more common in discrete parts manufacturing and batch process control than in continuous process control. In SPC, "process" refers to the work and quality control process, rather than continuous process control.

- Batch process control is employed in non-continuous batch processes, such as many pharmaceuticals, chemicals, and foods.

- Simulation-based optimization incorporates dynamic or steady-state computer-based process simulation models to determine more optimal operating targets in real-time, *i.e.* on a periodic basis, ranging from hourly to daily. This is sometimes considered a part of APC, but in practice it is still an emerging technology and is more often part of MPO.

- Manufacturing planning and optimization (MPO) refers to ongoing business activity to arrive at optimal operating targets that are then implemented in the operating organization, either manually or in some cases automatically communicated to the process control system.

- Safety instrumented system refers to a system that is independent of the process control system, both physically and administratively, whose purpose is to assure basic safety of the process.

APC Business and Professionals

Those responsible for the design, implementation and maintenance of APC applications are often referred to as APC Engineers or Control Application Engineers. Usually their education is dependent upon the field of specialization. For example, in the process industries many APC Engineers have a chemical engineering background, combining process control and chemical processing expertise.

Most large operating facilities, such as oil refineries, employ a number of control system specialists and professionals, ranging from field instrumentation, regulatory control system (DCS and PLC), advanced process control, and control system network and security. Depending on facility size and circumstances, these personnel may have responsibilities across multiple areas, or be dedicated to each area. There are also many process control service companies that can be hired for support and services in each area.

CASADE CONTROL

Applications with two or more capacities (such as heated jackets) are inherently difficult to control with a single control loop due to large overshoots and unacceptable lags. The solution is a cascade of two or more control loops, each with its own input, in series forming a single regulating device.

The product setpoint temperature is set on the master control loop. This is compared to the product temperature, and the master's PID output is used to set the remote setpoint of the slave. This is scaled to suit any expected temperature. The slave loop's natural response time should ideally be at least 5 times faster than the master.

Historically Cascade Control has commonly been achieved by using 2 or more individual controllers, however it is possible to achieve this using a dual or multi loop controller that offers cascade control functionality.

What is Cascade Control?

In single-loop control, the controller's set point is set by an operator, and its output drives a final control element. For example: a level controller driving a control valve to keep the level at its set point.

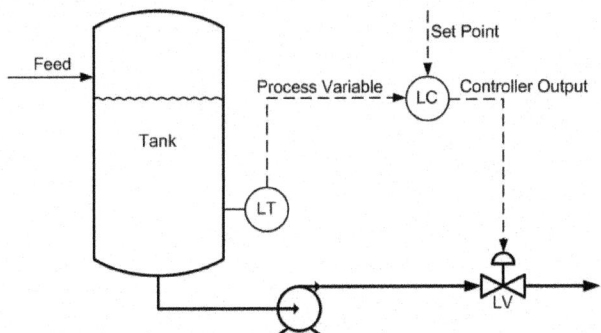

Fig. : Single Loop Control.

In a cascade control arrangement, there are two (or more) controllers of which one controller's output drives the set point of another controller. For example: a level controller driving the set point of a flow controller to keep the level at its set point. The flow controller, in turn, drives a control valve to match the flow with the set point the level controller is requesting.

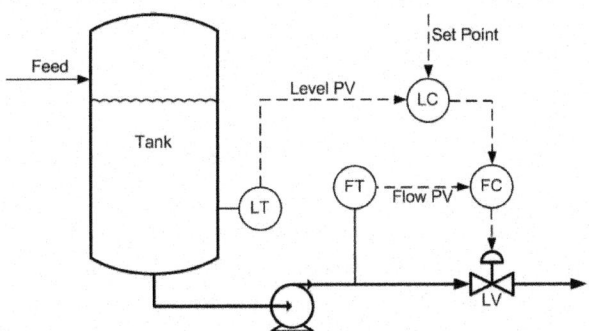

Fig. : Cascade Control.

The controller driving the set point (the level controller in the example above) is called the primary, outer, or master controller. The controller receiving the set point (flow controller in the example) is called the secondary, inner or slave controller.

What are the Advantages of Cascade Control?

There are several advantages of cascade control, and most of them boil down to isolating a slow control loop from nonlinearities in the final control element. In the example above the relatively slow level control loop is isolated from any control valve problems by having the fast flow control loop deal with these problems.

Imagine that the control valve has a stiction problem. Without the flow control loop, the level control loop (driving the sticky valve) will continuously oscillate in a stick-slip cycle with a long (slow) period, which will quite likely affect the downstream process. With the fast flow control loop in place, the sticky control valve will cause it to oscillate, but at a much shorter (faster) period due to the inherent fast dynamic behaviour of a well-tuned flow loop. It is likely that the fast oscillations will be attenuated by the downstream process without having much of an adverse effect.

Or imagine that the control valve has a nonlinear flow characteristic. This requires that the control loop driving it be detuned to maintain stability through-out the possible range of flow rates. (Of course there are better ways to deal with nonlinearities, but that is the topic of another blog.) If the level controller directly drives the valve, it must be detuned to maintain stability – possibly resulting in very poor level control. In a cascade control arrangement with a flow control loop driving the valve, the flow loop will be detuned to maintain stability. This will result in relatively poor flow control, but because the flow loop is dynamically so much faster than the level loop, the level control loop is hardly affected.

When Should Cascade Control be Used?

Cascade control should always be used if you have a process with relatively slow dynamics (like level, temperature, composition, humidity) and a liquid or gas flow, or some other relatively-fast process, has to be manipulated to control the slow process. For example: changing cooling water flow rate to control condenser pressure (vacuum), or changing steam flow rate to control heat exchanger outlet temperature. In both cases, flow control loops should be used as inner loops in cascade arrangements.

Does Cascade Control Have any Disadvantages?

Cascade control has three disadvantages. One, it requires an additional measurement (usually flow rate) to work. Two, there is an additional controller that has to be tuned. And three, the control strategy is more complex – for engineers and operators alike. These disadvantages have to be weighed up against the benefits of the expected improvement in control to decide if cascade control should be implemented.

When Should Cascade Control Not be Used?

Cascade control is beneficial only if the dynamics of the inner loop are fast compared to those of the outer loop. Cascade control should generally not be used if the inner loop is not at least three times faster than the outer loop, because the improved performance may not justify the added complexity.

In addition to the diminished benefits of cascade control when the inner loop is not significantly faster than the outer loop, there is also a risk of interaction between the two loops that could result in instability – especially if the inner loop is tuned very aggressively.

How Should Cascade Controls be Tuned?

A cascade arrangement should be tuned starting with the innermost loop. Once that one is tuned, it is placed in cascade control, or external set point mode, and then the loop driving its set point is tuned. Do not use quarter-amplitude-damping tuning rules (such as the unmodified Ziegler-Nichols and Cohen-Coon rules) to tune control loops in a cascade structure because it can cause instability if the process dynamics of the inner and outer loops are similar.

What is Meant by Cascade Control?

Cascade control can significantly improve the control quality. This applies in particular to the dynamic action of the control loop, in other words, the transition of the process variable following setpoint changes or disturbances.

Example 1: schematic construction of a cascade

Chocolate has to be heated to v_s = 40 °C for processing. The chocolate temperature must nowhere exceed 50 °C (even close to the heater). It is therefore heated on a water bath.

Cascade control is used in order to achieve rapid stabilisation.

Controller 1 is always the master controller, controller 2 always the slave.

The setpoint for the slave controller is produced by output conversion.

The control output y_1 is converted to a setpoint using the unit of the process value x_2 (here: 0 - 100 % = 0 - 50 °C).

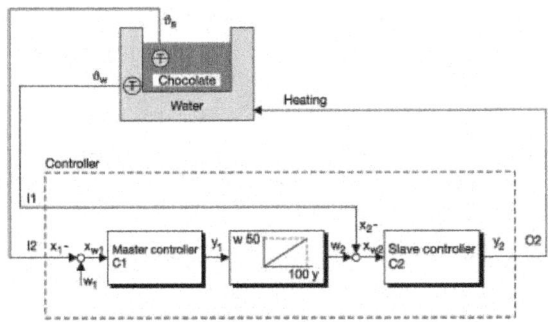

List of symbols

O2 - Output 2

I1 - Analogue input 1

I2 - Analogue input 2

C1 - Controller 1

C2 - Controller 2

w1 - Setpoint controller 1

w2 - Setpoint controller 2

x1 - Process value controller 1

x2 - Process value controller 2

xw1 - Deviation controller 1

xw2 - Deviation controller 2

y1 - Control output 1

y2 - Control output 2; output 1 of controller 2

vs - Chocolate temperature

vw - Water bath temperature

Example 2: construction of a trimming cascade

Two charges of chocolate have to be heated to 40 °C and 50 °C. The chocolate temperature must nowhere (not even close to a heater) exceed the setpoint by more than 10 °C. It is therefore heated on a water bath.

Trim cascade control is used to achieve rapid stabilisation without overshoot and without altering the controller configuration (output conversion) at a change of setpoint (batch change).

Controller 1 is always the master controller, controller 2 always the slave controller.

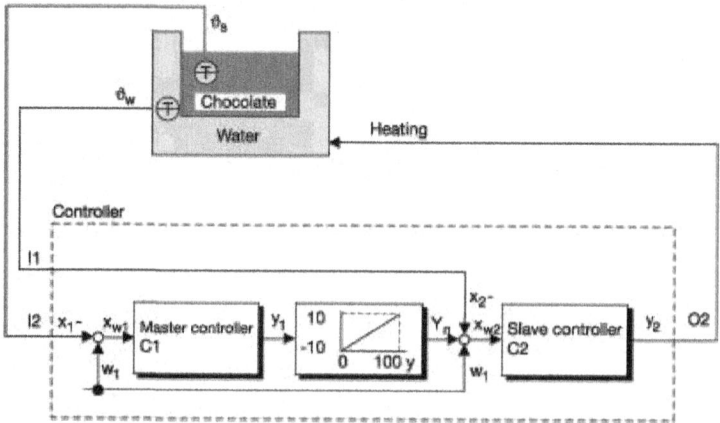

The setpoint for the slave controller is produced by output conversion and the addition of the master controller setpoint (w_1).

In setpoint conversion, the control output y_1 is converted to a value with the unit of the process value w_2. It corresponds to the maximum permitted temperature difference ($\pm \mid x_1 - w_1 \mid$; here: 0 - 100 % = -10 to +10 °C).

List of symbols

O2 - Output 2

I1 - Analogue input 1

I2 - Analogue input 2

C1 - Controller 1

C2 - Controller 2

w1 - Setpoint controller 1

x1 - Process value controller 1

x2 - Process value controller 2

xw1 - Deviation controller 1

xw2 - Deviation controller 2

y1 - Control output 1

y2 - Control output 2; output 1 o controller 2

vs - Chocolate temperature

vw - Water bath temperature

he Pro-EC44 Dual Loop Controller w/ Cascade Control Function - Learn More

Tuning for Cascade Control

First set the master to manual mode. Tune the slave control loop using proportional control only (I & D are not normally required) then return the master to automatic mode before tuning the master.

Example of a Cascade Control Application

In the example below a product temperature is being controlled via a heated oil jacket. The maximum input represents 400°C, thus restricting the jacket temperature.

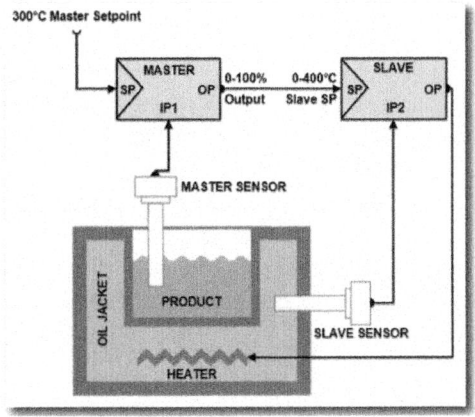

At start-up the master compares the product temperature (ambient) to its setpoint (300°C) and gives maximum output. This sets the maximum (400°C) setpoint on the slave, which is compared to the jacket temperature (ambient) giving maximum heater output. As the jacket temperature rises towards setpoint, the slave's heater output falls. The product's temperature will also have begun rising at a rate dependant on the transfer lag between the jacket and product. This causes the masters PID output to decrease, reducing the 'jacket' setpoint on the slave, effectively reducing the output to the heater. This continues until the system becomes balanced.

The result is quicker, smoother control with minimum overshoot and the ability to cope with changes in the load, whilst keeping the jacket temperature within acceptable tolerances.

THE CASCADE CONTROL ARCHITECTURE

Two popular control strategies for improved disturbance rejection performance are cascade control and feed forward with feedback trim.

Improved performance comes at a price. Both strategies require that additional instrumentation be purchased, installed and maintained. Both also require additional engineering time for strategy design, tuning and implementation.

The cascade architecture offers alluring additional benefits such as the ability to address multiple disturbances to our process and to improve set point response performance.

In contrast, the feed forward with feedback trim architecture is designed to address a single measured disturbance and does not impact set point response performance in any fashion.

The Inner Secondary Loop

The dashed line in the block diagram below circles a **feedback control loop** like we have discussed. The only difference is that the words "inner secondary" have been added to the block descriptions. The variable labels also have a "2" after them.

Traditional Feedback Control Loop is in the Dashed Circle

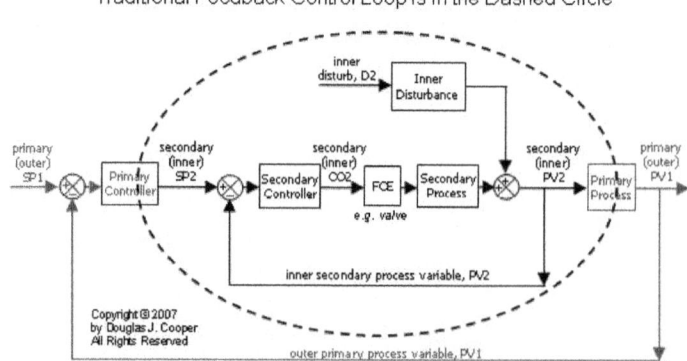

So,

SP2 = inner secondary set point

CO2 = inner secondary controller output signal

PV2 = inner secondary measured process variable signal

And

D2 = inner disturbance variable (often not measured or available as a signal)

FCE = final control element such as a valve, variable speed pump or compressor, *etc.*

The Nested Cascade Architecture

To construct a cascade architecture, we literally nest the secondary control loop inside a primary loop as shown in the block diagram below.

Note that outer primary PV1 is our process variable of interest in this implementation. PV1 is the variable we would be measuring and controlling if we had chosen a traditional single loop architecture instead of a cascade.

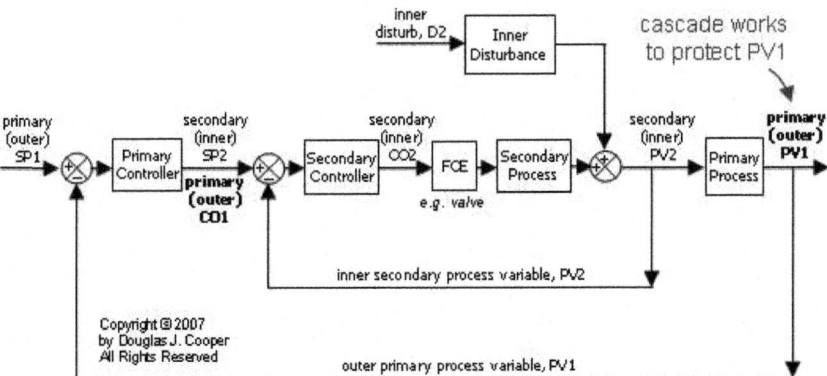

Cascade Structure is a Control Loop within a Control Loop

Because we are willing to invest the additional effort and expense to improve the performance response of PV1, it is reasonable to assume that it is a variable important to process safety and/or profitability. Otherwise, it does not make sense to add the complexity of a cascade structure.

Naming Conventions

Like many things in the PID control world, vendor documentation is not consistent. The most common naming conventions we see for cascade (also called nested) loops are:

- secondary and primary
- inner and outer
- slave and master

In an attempt at clarity, we are somewhat repetitive by using labels like "inner secondary" and "outer primary."

Two PVs, Two Controllers, One Valve

Notice from the block diagrams that the cascade architecture has:

- two controllers (an inner secondary and outer primary controller)
- two measured process variable sensors (an inner PV2 and outer PV1)
- only *one* final control element (FCE) such as a valve, pump or compressor.

How can we have two controllers but only one FCE? Because as shown in the diagram above, the controller output signal from the outer primary controller, CO1, becomes the set point of the inner secondary controller, SP2.

The outer loop literally commands the inner loop by adjusting its set point. Functionally, the controllers are wired such that SP2 = CO1.

This is actually good news from an implementation viewpoint. If we can install and maintain an inner secondary sensor at reasonable cost, and if we are using a **PLC** or **DCS** where adding a controller is largely a software selection, then the task of constructing a cascade control structure may be reasonably straightforward.

Early Warning is Basis for Success

An essential element for success in a cascade design is the measurement and control of an "early warning" process variable.

Cascade Control Depends on an Inner "Early Warning" Variable

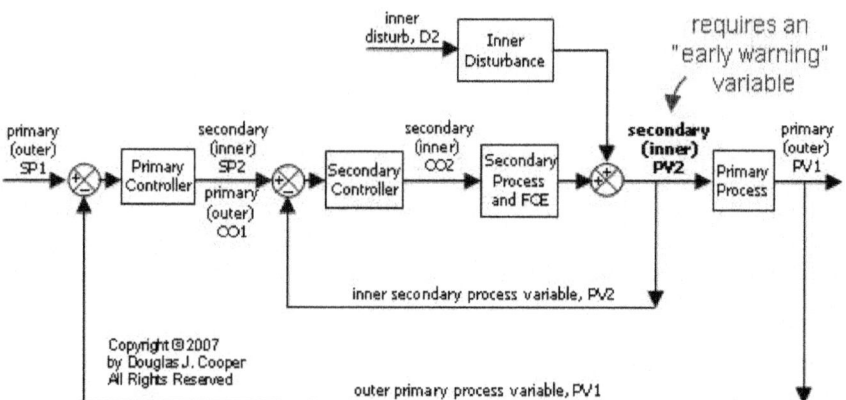

In the cascade architecture, inner secondary PV2 serves as this early warning process variable. Given this, essential design characteristics for selecting PV2 include that:

- it be measurable with a sensor,
- the same FCE (*e.g.*, valve) used to manipulate PV1 also manipulates PV2,

- the same disturbances that are of concern for PV1 also disrupt PV2, and

- PV2 responds *before* PV1 to disturbances of concern and to FCE manipulations.

Since PV2 sees the disruption first, it provides our "early warning" that a disturbance has occurred and is heading toward PV1. The inner secondary controller can begin corrective action immediately. And since PV2 responds first to final control element (*e.g.*, valve) manipulations, disturbance rejection can be well underway even before primary variable PV1 has been substantially impacted by the disturbance.

With such a cascade architecture, the control of the outer primary process variable PV1 benefits from the corrective actions applied to the upstream early warning measurement PV2.

Disturbance Must Impact Early Warning Variable PV2

The even with a cascade structure, there will likely be disturbances that impact PV1 but do not impact early warning variable PV2.

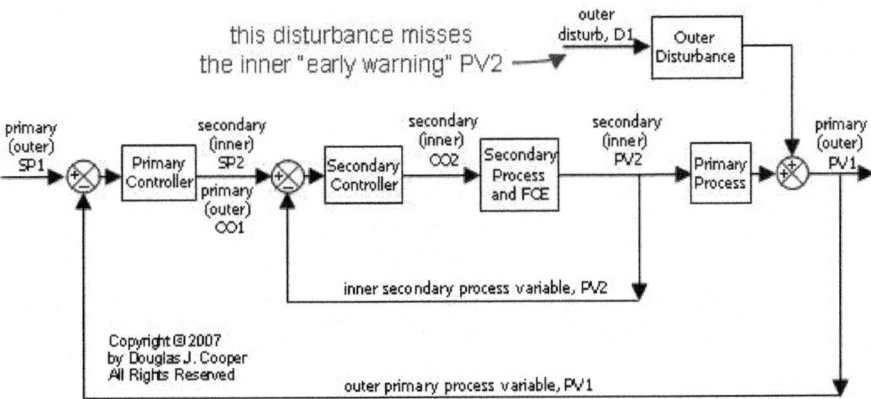

Disturbance Must Hit the Inner PV for Cascade to Provide Benefit

The inner secondary controller offers no "early action" benefit for these outer disturbances. They are ultimately addressed by the outer primary controller as the disturbance moves PV1 from set point.

On a positive note, a proper cascade can improve rejection performance for any of a host of disturbances that directly impact PV2 before disrupting PV1.

An Illustrative Example

To illustrate the construction and value of a cascade architecture, consider the liquid level control process shown below. This is a variation on our **gravity drained tanks**, so hopefully, the behaviour of the process below follows intuitively from our previous investigations.

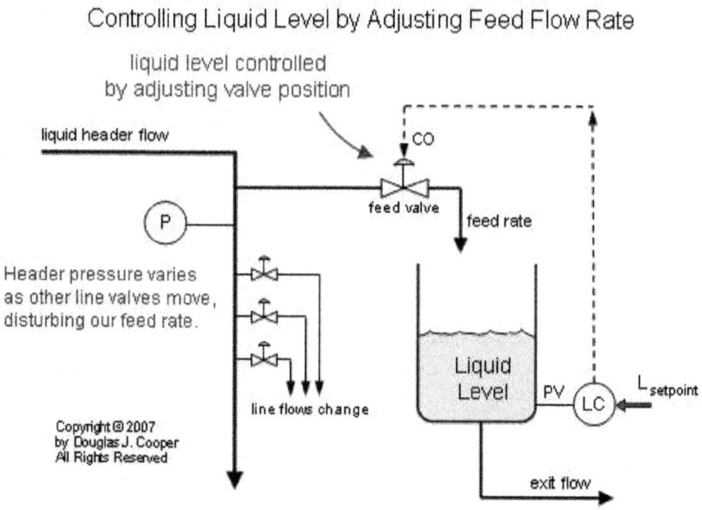

Controlling Liquid Level by Adjusting Feed Flow Rate

liquid level controlled
by adjusting valve position

liquid header flow

CO

feed valve

feed rate

P

Header pressure varies
as other line valves move,
disturbing our feed rate.

Liquid
Level

PV

LC

setpoint

line flows change

exit flow

The tank is essentially a barrel with a hole punched in the bottom. Liquid enters through a feed valve at the top of the tank. The exit flow is liquid draining freely by the force of gravity out through the hole in the tank bottom.

The control objective is to maintain liquid level at set point (SP) in spite of unmeasured disturbances. Given this objective, our measured process variable (PV) is liquid level in the tank. We measure level with a sensor and transmit the signal to a level controller (the LC inside the circle in the diagram).

After comparing set point to measurement, the level controller (LC) computes and transmits a controller output (CO) signal to the feed valve. As the feed valve opens and closes, the liquid feed rate entering the top of the tank increases and decreases to raise and lower the liquid level in the tank.

This "measure, compute and act" procedure repeats every loop sample time, T, as the controller works to maintain tank level at set point.

The Disturbance

The disturbance of concern is the pressure in the main liquid header. The header supplies the liquid that feeds our tank. It also supplies liquid to several other lines flowing to different process units in the plant.

Whenever the flow rate of one of these other lines changes, the header pressure can be impacted. If several line valves from the main header open at about the same time, for example, the header pressure will drop until its own control system corrects the imbalance. If one of the line valves shuts in an emergency action, the header pressure will momentarily spike.

As the plant moves through the cycles and fluctuations of daily production, the header pressure rises and falls in an unpredictable fashion. And every time the header pressure changes, the feed rate to our tank is impacted.

Problem with Single Loop Control

The single loop architecture attempts to achieve our control objective by adjusting valve position in the liquid feed line. If the measured level is higher than set point, the controller signals the valve to close by an appropriate percentage with the expectation that this will decrease feed flow rate accordingly.

But feed flow rate is a function of two variables:

- feed valve position, and
- the header pressure pushing the liquid through the valve (a disturbance).

To explore this, we conduct some thought experiments:

Thought Experiment #1: Assume that the main header pressure is perfectly constant over time. As the feed valve opens and closes, the feed flow rate and thus tank level increases and decreases in a predictable fashion. In this case, a single loop structure provides acceptable level control performance.

Thought Experiment #2: Assume that our feed valve is set in a fixed position and the header pressure starts rising. Just like squeezing harder on a spray bottle, the valve position can remain constant yet the rising pressure will cause the flow rate through the fixed valve opening to increase.

*Thought Experiment #3: Now assume that the header pressure starts to rise at the same moment that the controller determines that the liquid level in our tank is too high. The controller can be **closing** the feed valve, but because header pressure is rising, the flow rate through the valve can actually be **increasing**.*

As presented in Thought Experiment #3, The changing header pressure (a disturbance) can cause a contradictory outcome that can confound the controller and degrade control performance.

A Cascade Control Solution

For high performance disturbance rejection, it is not valve position, but rather, feed flow rate that must be adjusted to control liquid level.

Because header pressure changes, increasing feed flow rate by a precise amount can sometimes mean opening the valve a lot, opening it a little, and because of the changing header pressure, perhaps even closing the valve a bit.

Below is a classic level-to-flow cascade architecture. As shown, an inner secondary sensor measures the feed flow rate. An inner secondary controller receives this flow measurement and adjusts the feed flow valve.

With this cascade structure, if liquid level is too high, the primary level controller now calls for a decreased liquid feed flow rate rather than simply a decrease in valve opening. The flow controller then decides whether this means opening or closing the valve and by how much.

Note in the diagram that, true to a cascade, the level controller output signal (CO1) becomes the set point for the flow controller (SP2).

Level-to-Flow Cascade Control

liquid level controlled
by adjusting flow rate

$F_{setpoint}$ CO1 =SP2

primary liquid header flow   FC ---- CO2

PV2

P

disturbance
to feed rate

feed rate

line flows change

Liquid
Level   PV1   LC $L_{setpoint}$

exit flow

Header pressure disturbances are quickly detected and addressed by the secondary flow controller. This minimizes any disruption caused by changing header pressure to the benefit of our primary level control process.

The Level-to-Flow Cascade Block Diagram

As shown in the block diagram below, our level-to-flow cascade fits into our block diagram structure. As required, there are:

- Two controllers – the outer primary level controller (LC) and inner secondary feed flow controller (FC)

- Two measured process variable sensors – the outer primary liquid level (PV1) and inner secondary feed flow rate (PV2)

- One final control element (FCE) – the valve in the liquid feed stream

Formal Level-to-Flow Cascade Structure

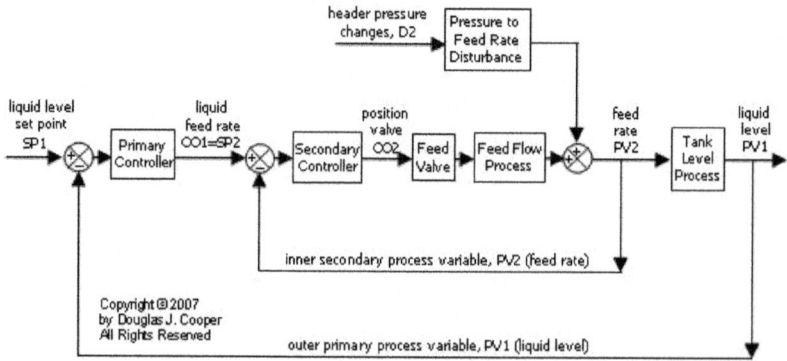

header pressure
changes, D2   Pressure to
Feed Rate
Disturbance

liquid level
set point
SP1

Primary
Controller

liquid
feed rate
CO1=SP2

Secondary
Controller

position
valve
CO2

Feed
Valve

Feed Flow
Process

feed
rate
PV2

Tank
Level
Process

liquid
level
PV1

inner secondary process variable, PV2 (feed rate)

outer primary process variable, PV1 (liquid level)

As required for a successful design, the inner secondary flow control loop is nested inside the primary outer level control loop. That is:

- The feed flow rate (PV2) responds before the tank level (PV1) when the header pressure disturbs the process or when the feed valve moves.
- The output of the primary controller, CO1, is wired such that it becomes the set point of the secondary controller, SP2.
- Ultimately, level measurement, PV1, is our process variable of primary concern. Protecting PV1 from header pressure disturbances is the goal of the cascade.

Design and Tuning

The inner secondary and outer primary controllers are from the PID family of algorithms. We have explored the design and tuning of these controllers in numerous, so as we will see, implementing a cascade builds on many familiar tasks.

FUNDAMENTALS OF CASCADE CONTROL

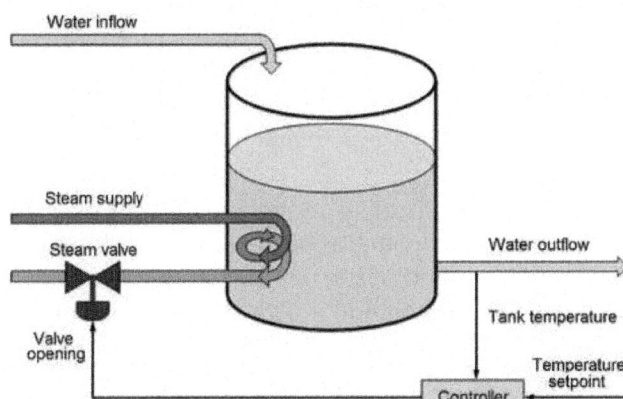

When multiple sensors are available for measuring conditions in a controlled process, a cascade control system can often perform better than a traditional single-measurement controller. Consider, for example, the steam-fed water heater shown in the sidebar Heating Water with Cascade Control. A traditional controller is shown measuring the temperature inside the tank and manipulating the steam valve opening to add more or less heat as inflowing water disturbs the tank temperature. This arrangement works well enough if the steam supply and the steam valve are sufficiently consistent to produce another X% change in tank temperature every time the controller calls for another Y% change in the valve opening.

However, several factors could alter the ratio of X to Y or the time required for the tank temperature to change after a control effort. The pressure in the steam supply line could drop while other tanks are drawing down the steam supply they share, in which case the controller would have to open the valve more than Y% in order to achieve the same X% change in tank temperature.

Or, the steam valve could start sticking as friction takes its mechanical toll over time. That would lengthen the time required for the valve to open to the extent called for by the controller and slow the rate at which the tank temperature changes in response to a given control effort.

A Better Way

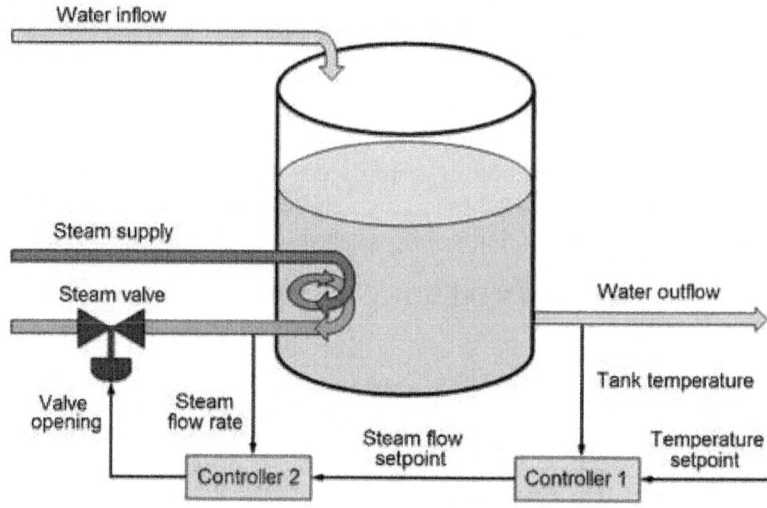

A cascade control system could solve both of these problems where a second controller has taken over responsibility for manipulating the valve opening based on measurements from a second sensor monitoring the steam flow rate. Instead of dictating how widely the valve should be opened, the first controller now tells the second controller how much heat it wants in terms of a desired steam flow rate.

The second controller then manipulates the valve opening until the steam is flowing at the requested rate. If that rate turns out to be insufficient to produce the desired tank temperature, the first controller can call for a higher flow rate, thereby inducing the second controller to provide more steam and more heat (or vice versa).

That may sound like a convoluted way to achieve the same result as the first controller could achieve on its own, but a cascade control system should be able to provide much faster compensation when the steam flow is disturbed. In the original single-controller arrangement, a drop in the steam supply pressure would first have to lower the tank temperature before the temperature sensor could even notice the disturbance. With the second controller and second sensor on the job, the steam flow rate can be measured and maintained much more quickly and precisely, allowing the first controller to work with the belief that whatever steam flow rate it wants it will in fact get, no matter what happens to the steam pressure.

The second controller can also shield the first controller from deteriorating valve performance. The valve might still slow down as it wears out or gums up, and the second controller might have to work harder as a result, but the first controller would be unaffected as long as the second controller is able to maintain the steam flow rate at the required level.

Without the acceleration afforded by the second controller, the first controller would see the process becoming slower and slower. It might still be able to achieve the desired tank temperature on its own, but unless a perceptive operator notices the effect and re-tunes it to be more aggressive about responding to disturbances in the tank temperature, it too would become slower and slower.

Similarly, the second controller can smooth out any quirks or non-linearities in the valve's performance, such as an orifice that is harder to close than to open. The second controller might have to struggle a bit to achieve the desired steam flow rate, but if it can do so quickly enough, the first controller will never see the effects of the valve's quirky behaviour.

Elements of Cascade Control

The Cascade Control Block Diagram shows a generic cascade control system with two controllers, two sensors, and one actuator acting on two processes in series. A primary or master controller generates a control effort that serves as the setpoint for a secondary or slave controller. That controller in turn uses the actuator to apply its control effort directly to the secondary process. The secondary process then generates a secondary process variable that serves as the control effort for the primary process.

The geometry of this block diagram defines an inner loop involving the secondary controller and an outer loop involving the primary controller. The inner loop functions like a traditional feedback control system with a setpoint, a process variable, and a controller acting on a process by means of an actuator. The outer loop does the same except that it uses the entire inner loop as its actuator.

In the water heater example, the tank temperature controller would be primary since it defines the setpoint that the steam flow controller is required to achieve. The water in the tank, the tank temperature, the steam, and the steam flow rate would be the primary process, the primary process variable, the secondary process, and the secondary process variable, respectively (refer to the Cascade Control Block Diagram). The valve that the steam flow controller uses to maintain the steam flow rate serves as the actuator which acts directly on the secondary process and indirectly on the primary process.

Requirements

Naturally, a cascade control system can't solve every feedback control problem, but it can prove advantageous if under the right circumstances:

- The inner loop has influence over the outer loop. The actions of the secondary controller must affect the primary process variable in a predictable and repeatable way or else the primary controller will have no mechanism for influencing its own process.

- The inner loop is faster than the outer loop. The secondary process must react to the secondary controller's efforts at least three or four times faster than the primary process reacts to the primary controller. This allows the secondary controller enough time to compensate for inner loop disturbances before they can affect the primary process.

- The inner loop disturbances are less severe than the outer loop disturbances. Otherwise, the secondary controller will be constantly correcting for disturbances to the secondary process and unable to apply consistent corrective efforts to the primary process.

Steam-fed water heaters as in the example are particularly amenable to cascade control because raising or lowering the steam flow rate raises or lowers the tank temperature without any additional actuators, a valve can manipulate a steam flow rate almost instantaneously in comparison to the slow pace at which steam can heat the water in a large tank, and disturbances to the steam supply pressure are relatively infrequent and easily compensated by the steam flow controller.

Cascade Control Block Diagram

A cascade control system reacts to physical phenomena shown in blue and process data shown in green.

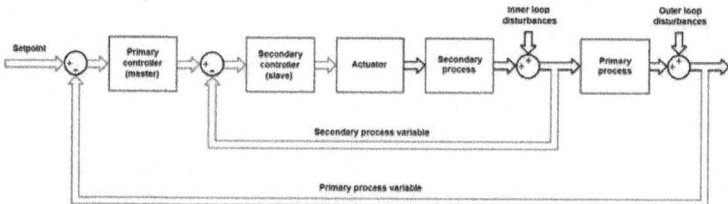

In the water heater example:
- Setpoint - temperature desired for the water in the tank
- Primary controller (master) - measures water temperature in the tank and asks the secondary controller for more or less heat
- Secondary controller (slave) - measures and maintains steam flow rate directly
- Actuator - steam flow valve
- Secondary process - steam in the supply line
- Inner loop disturbances - fluctuations in steam supply pressure
- Primary process - water in the tank
- Outer loop disturbances - fluctuations in the tank temperature due to uncontrolled ambient conditions, especially fluctuations in the inflow temperature
- Secondary process variable - steam flow rate
- Primary process variable - tank water temperature

Challenges

Cascade control can also have its drawbacks. Most notably, the extra sensor and controller tend to increase the overall equipment costs. Cascade control systems are also more complex than single-measurement controllers, requiring twice as much tuning. Then again, the tuning procedure is fairly straightforward: tune the secondary controller first, then the primary controller using the same tuning tools applicable to single-measurement controllers.

However, if the inner loop tuning is too aggressive and the two processes operate on similar time scales, the two controllers might compete with each other to the point of driving the closed-loop system unstable. Fortunately, this is unlikely if the inner loop is inherently faster than the outer loop or the tuning forces it to be.

And it's not always clear when cascade control will be worth the extra effort and expense. There are several classic examples that typically benefit from cascade control-often involving a flow rate as the secondary process variable-but it's usually easier to predict when a cascade control system won't help than to predict when it will.

REGULATORY CONTROL IS THE FOUNDATION FOR ADVANCED PROCESS CONTROL

W ant top-notch advanced process control? Get the basic regulatory control right, first.

A dictionary definition of optimization might read: 'A strategy giving the best result obtainable under a given set of conditions.' To the practicing control engineer, optimization usually means a highly theoretical exercise, which is not really relevant in the real world, where pipes leak, sensors plug, pumps cavitate, and valves stick.

Optimization in an operating plant requires integrating process know-how to maximize productivity, including paying attention to field devices, control strategy suitability, controller tuning, and wise use of basic regulatory control (BRC)-cascade, feedback, feedforward, ratio, lead/lag, *etc.*-at the lowest possible level of implementation. It is upon this foundation that effective advanced process control (APC) can be constructed.

Maximizing return on investment (ROI) is always an important goal of plant management. However, purchasing control equipment, transmitters, and a digital control system loaded with the latest advanced control software does not guarantee good control. If not managed, understood, and optimized to match control objectives for which these purchases were made, the ROI is poor.

Function of Control Systems

Process control systems convert raw materials and energy into usable products through an intricate series of processing steps and control techniques. By themselves, plant processes do not produce levels, flows, or temperatures and the introduction of such variables represent-operating constraints designed to ensure productivity, efficiency, and/or quality product is produced.

For a plant to produce high quality product at the least costs requires a structure of people, systems, and facilities integrated together and built on a foundation of management commitment.

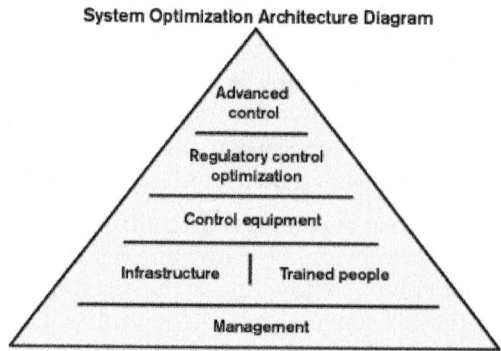

System Optimization Architecture Diagram

Among the most significant management challenges are cost containment, product quality, conforming to government regulations, and achievement of production schedule targets. Managers who recognize and embrace the value robust process control contributes toward overall operating results gain competitive advantage for their company. However, many managers have been 'burned' by past promises of improved process performance from control system investments that never materialized. Too frequently, the result has been underutilized control systems.

Gaining maximum advantage from the available capability of currently installed control systems requires an organizational philosophy that empowers, supports, and organizes in a way that achieves the desired results. Simply purchasing advanced hardware and software does not ensure success; without management's commitment any system optimization program is doomed to failure. Success requires effective management of the people, facilities, and systems working together to achieve established goals.

People and Systems

For any optimization process to be successful, an infrastructure of systems must be in place that provide the training and tools so qualified people can successfully carry out activities that lead to meeting defined goals and objectives. The introduction of high technology equipment-digital control systems, graphic monitors, smart transmitters, and artificial intelligence-into the work place do not reduce, but increase the requirement for training personnel.

Effectively optimizing regulatory functions of a process control system requires knowledge of the static and dynamic factors that exist between, and among, loops.

Control theory, as presented in university courses, is typically abstract mathematics. Once out of school, control engineers quickly forget what was learned and process control is practiced on an *ad hoc* basis without reference to the theory that governs the behaviour of dynamic systems.

During the past sixteen years, Techmation (Scottsdale, Ariz.) consultants have worked in over 2,700 operating plants, testing, analyzing, and improving the dynamic operational characteristics of tens of thousands of control loops. During this time procedures and data analysis techniques have been developed and refined into a knowledge base that bridges the gap between theory and real-world practice.

Because process operations are dynamic, it's important that after consultants complete and leave, customer employees can maintain the control system at the level of optimization achieved.

To ensure a satisfactory level of customer employee proficiency, consultants mix classroom theory with practical experience to form a process audit and improvement team.

Following classroom instruction on how to efficiently and effectively use a personal computer to test and analyze process data, the audit team, under the supervision of the consultant, apply a methodology of identifying, auditing, testing, analyzing, and optimizing to establish a solid foundation for APC. Techmation's methodology follows.

- Indentification consists of designing and conducting tests that identify critical measurements and control loops.

- Auditing uses test data to determine the accuracy and capability of all measurements and loops. The team corrects problems discovered during the audit, such as calibration and valve errors. Critical loops are tested to determine installed characteristic. When installed characteristic are found to be highly non-linear, especially in the normal operating range, the problem is corrected.

Where required, controller algorithms are re-programmed and re-tuned to meet control requirements. If audit testing indicates installed control strategies do not meet objectives, the audit team designs, implements, tests, and documents new strategies that will meet defined control objectives. When completed, the audit report provides a 'capability signature' of the process personnel can use to maintain system integrity

- Analyzing and optimizing consists of using new data, collected following the audit phase, to identify where changes or new control strategies could improve unit control. During this phase, the emphasis is on identifying unit optimization strategies and system configuration implementation methods.

By following an identifying, auditing, testing, analyzing, and optimizing methodology and mixing formal classroom training with practical experience, under the guidance of a consultant, users can optimize regulatory control systems and ensure they stay optimized.

Process Control Equipment, Strategies

A poorly designed control strategy, inaccurate measurements, and poorly functioning control valves will result in less than optimum control.

Control system testing typically documents a number of equipment related problems that need to be fixed before the regulatory control systems dynamic operation can be optimized. Attempting to mask equipment problems by adjusting the PID (proportional/integral/derivative) and filter settings in controllers is often the response where issues of training, time constraints, and management commitment are manifest. In fact, as many as 50% of loops need some maintenance or configuration changes before the loop can be tuned to provide minimum variance regulatory control.

Regulatory control consists of two types of control functions that can be combined in an almost infinite number of configurations. The two types of regulatory control are: feedback and feedforward.

Feedback control can be configured for cascade, selective, ratio, and any number of other types of control schemes. All feedback control implementations have one thing in common. The controlled process variable measurement is compared to a reference, called the setpoint, and the deviation results in a corrective action by the controller. In short, feedback control only allows upsets to be corrected *after* they are detected. Tuning regulatory controls requires knowledge of the transfer function that describes the dynamic relationship between a change in the controller output and the response of the variable being controlled. Simplified first order models, rule of thumb procedures, and simplified rule based self-tuning controllers can, not in many cases, provide the best regulatory control solution for many of the dynamically complex processes found in control applications.

Feedforward control measures a load disturbance and introduces a dynamically compensated corrective action before the load disturbance affects the controlled variable. Tuning of feedforward control loops requires knowledge of the process transfer function in addition to the load upset transfer function. The only way to accurately obtain the necessary knowledge of the system dynamics is to test the installed system under operating conditions to obtain time domain data that accurately represents the process response. Time domain data must be accurately transformed into the frequency domain to obtain the best control solutions.

The function of the regulatory portion of the control system is to reduce variability in the face of changing conditions. Without an effective regulatory control system, each successive unit operation can introduce variation that can accumulate throughout the process and is reflected in the final product quality and overall cost of production. To produce a uniform product that consistently meets customer demands at the lowest cost, a regulatory control system must be in place to minimize variance throughout the processing cycle.

A modern process plant may have thousands of control loops. Advanced control systems cannot be operated as designed without a majority of these loops being in automatic control. The regulatory control loops provide four functions:

1. Allow the process to operate at a chosen target;

2. Minimize effects of load disturbances;

3. Reduce the effect of raw material variability; and

4. Provide for safe and efficient startup, operation, and shutdown of the process.

Hence, the regulatory control system's function is to maximize product uniformity under dynamic conditions.

Regulatory Controllers

PID is the most common feedback controller, has been in use for more than 60 years, yet is not covered by any implementation standard. In fact, no two digital control systems implement PID control in the same manner, understandably complicating tuning methods and calculations. For example, digital controller manufactures write software for PID controllers using the ideal, series, or parallel algorithm configurations. Because of PID implementation algorithm differences, loop-tuning parameters can be significantly different to accomplish the same task depending on the algorithm type implemented. Adding to the confusion, controller-tuning units can be expressed in different units and time domains such as:

- Proportional setting being in gain, percent proportional band, or throttling range;

- Integral setting in seconds per repeat, repeats per second, minutes per repeat, repeats per minute, or-in some cases-scan rate per repeat and repeats per scan rate; and

- Derivative terms expressed in seconds, minutes, or scan rate.

Muddying the waters still further, some manufactures allow the end-user to select the controller algorithm and tuning units. As if the preceding was not enough to create a sufficient tuning maze, add derivative filter constant settings, positional or velocity digital implementation of the PID, configurations for PID, PI-D, I-PD setpoint response, and PV filtering options, and it's easy to understand why 'simple' tuning can be confusing.

Conversely, over the years, numerous special linear and non-linear versions of the PID controller have been developed that provide better control of particular processes. For example, a non-linear PID controller algorithm is available that eliminates the stick-slip cycling found in as many as 30% of fast control loops.

Other special controller algorithms are used for averaging control in level systems, eliminating hysteresis cycling in integrating processes, preventing overshoot when filtering is used, and the conditional integral configuration for batch control to name just a few.

Understanding how to correctly implement these feedback controller algorithms is important to insure minimum variance control. Indeed, the crucial consideration in regulatory controllers is to understand the complex picture that exists with a trained eye toward the application of the appropriate techniques and solutions.

Testing Regulatory Loops

Attempts to achieve 'total product quality,' using SPC (statistical process control) and compliance with ISO quality standards, ignore control loop details such as:

- Correctly pairing of controlled and influential measurements;
- Hysteresis, stick-slip, and sizing of control valves; and
- Quality of the measurements, signal ailiasing problems, control algorithms, signal filtering, system configuration, and tuning of the regulatory control system.

Audit experience repeatedly indicates regulatory control systems are operational, but not providing optimum control. Findings indicate the typical regulatory control system contributes to as much as 50% of the non-uniformity of the final product. Testing of tens of thousands of unit operations with regulatory control systems applied consistently reveals the installed dynamics and loop tuning information, identifies equipment problems, installed characteristics of process loops, and relative gain of coupled loops.

Among the variety of problems identified, a few appear over and over in one of three general categories of: control valves, measurement, and control strategies.

Control Valve Problems

Stick-slip cycling -Tests reveal as many as 30% of rotary and high friction globe valves exhibit a tendency to produce stick-slip cycle at steady state operating conditions when the controller is tuned based on installed loop dynamics.

Typical Closed-loop Stick-slip Cycling Diagram

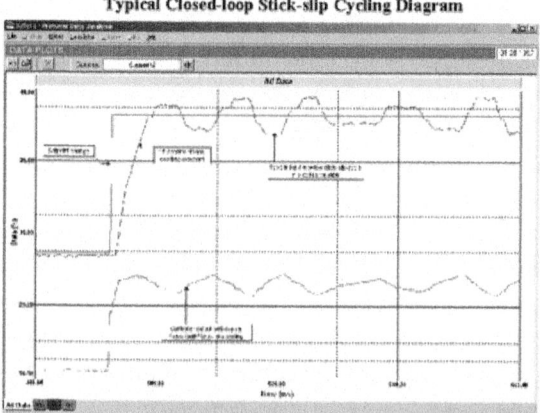

The stick-slip characteristics in a control valve results in cycling at steady state and can produce excessive process variability in unit operations. Stick-slip is a result of an excessive ratio of static to dynamic friction in the control valve, pneumatic stiffness in the actuator, and the performance of the valve positioner.

The steady state cycling of the controller output and the process variable being controlled is a typical stick-slip cycle. A linear PID controller measures the error between the setpoint and the process variable and ramps the controller output at a rate that is a function of the controller tuning parameters to correct the error. After a small and typically slow ramp change in the controller output, the valve position 'jumps' to a new position and the flow overshoots the setpoint. The error being on the other side of the setpoint causes the controller to again ramp its output to correct the error. The valve again 'jumps' to a new position and the cycle is repeated.

High-friction valves typically require tuning parameters that are 10 to 20 times slower than required, based on the installed loop dynamics to eliminate stick-slip cycling.

Correcting this very common problem requires a non-linear PID algorithm that sets the controller integral value to a lesser (slower) value only when the error is very small in the range where stick-slip cycling occurs. This algorithm has been installed in thousands of loops on numerous different controller brands with a net result of rejecting fast load disturbances and 'smoothing' steady state operation.

Hysteresis -Loose linkages in the actuator or positioner mountings-combined with friction in the valve-cause hysteresis or deadband in pneumatic control valves. Loop analysis testing reveals the normalized magnitude of the hysteresis in each loop.

Open Loop Hysteresis Check Diagram

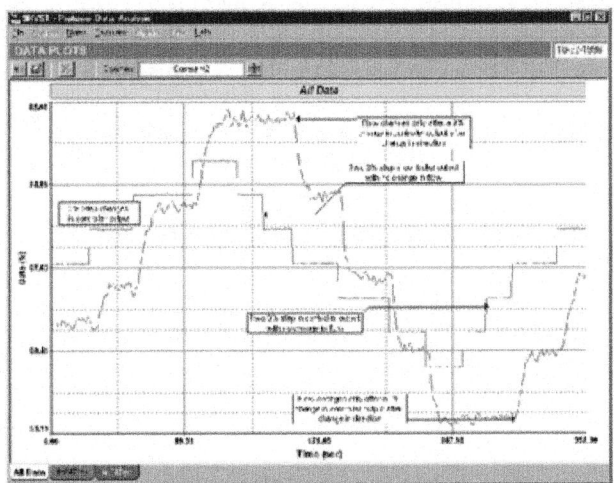

Small amounts of hysteresis can usually be tolerated in self-regulating loops but will result in continuous cycling at steady state in integrating loops such as level, large volume pneumatic pressure, and batch temperature loops.

Typical Closed Loop Hysteresis Cycle Under PI Control Diagram

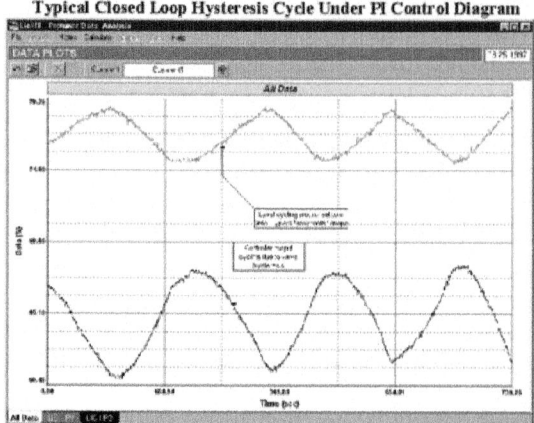

Techniques to eliminate hysteresis cycling in integrating loops include:

- Fixing the control valve;
- Placing the integrating loop in cascade where the inner loop is a self-regulating process; or
- Implementing an error squared on integral control algorithm.

Installed characteristic -No control loop has a completely linear installed characteristic. In most loops this non-linear installed characteristic can be easily handled using an appropriate gain margin in the calculation of the controller tuning parameters. In some instances the installed characteristic in the loop is so non-linear the loop must be made linear before the loop can be tuned for optimum closed loop response under all system load conditions. When this is the case software, such as Techmation's Protuner, is used to record the controller output in percent and the measured variable in percent. This data can be analyzed and used to determine the cam characteristic that will result in a linear response.

Installed Characteristic and Cam Equation Diagram

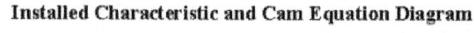

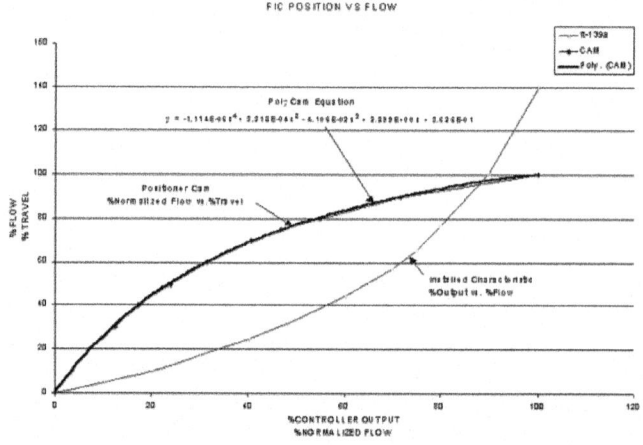

The illustration shows an equal percentage installed characteristic. Contrasted with linear characteristics, when the loop in the illustration is tuned with the valve at low-end travels, the loop will become unstable at high-end travels. Conversely, tuning the loop at high-end valve travel will cause sluggish performance at low-end valve travels. Even when the loop is tuned for stability under all loads, the non-linear installed characteristic will result in a varying closed-loop time constant. To operate properly, the APC closed loop model needs to be varied as a function of load.

The compensating cam is the mirror image of the normalized installed characteristic. The illustration indicates the cam characteristics graphically and as a polynomial equation, making it easy to implement in the controller output or as a digital cam in a digital (smart) positioner. Implementing the can will linearize the process and loop tuning will be effective under all load conditions with the added benefit of constant closed loop dynamic in support of APC modeling.

Other problems -Valve calibration and valve sizing are also common problems found.

A recent paper released by a major valve manufacture revealed as many as seventy percent of installed valves tested require zero and span calibration.

Incorrect zero and span of control valves can lead to various control problems not easily solved. For example, if the valve is calibrated to operate 0% to 100% travel from 10% to 80% controller outputs there is a potential integral windup problem. Windup can result in large and unexpected overshoot and slow startup of the affected variable.

Techmation's experience reveals as many as 30% of valves are oversized and 15% are undersized for the installed application. Oversized valves typically provide poor performance due to lack of rangibility; undersized valves can result in production bottlenecks.

Measurement Problems

A generally understood first law among process control engineers is, 'You can't control what you can't measure.'

In testing of regulatory control systems, impediments to accurately measure the variables being controlled is hindered by such things as:

- Lack of or improper setting of the required anti-aliasing filter constant in the transmitter;
- Excessive noise on the measurement signal;
- Improper use of or excessive signal filtering;
- Incorrect PID controller configurations when measurement filtering is present;
- Incorrect placement or mounting of measurement sensors; and
- Incorrect transmitter calibration and scaling.

Unless measurement issues are addressed, optimum regulatory control is impossible and APC cannot provide the increased system performance expected.

Control Strategy Problems

Experience indicates as many as 17% of regulatory unit operation control strategies are incorrectly implemented and must be redesigned. The following example illustrates the results of a typical control strategy redesign to control column pressure.

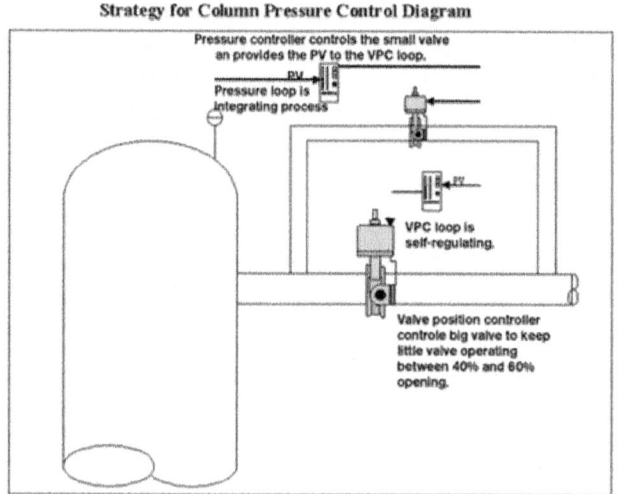

Strategy for Column Pressure Control Diagram

The original strategy used a single controller split ranged to control both valves. Testing the system revealed the pressure control loop was very non-linear using the original strategy. Because the process is a mass balance its dynamics are integrating. As shown in the 'Typical closed loop hysteresis cycle in level loop under PI control', any hysteresis or deadband in the valve will result in continuous cycling at steady state. To retain the original control strategy, and make it effective, would require linearizing the loop in software, and replacing the large butterfly valve because it is not a precision control valve.

The new strategy retains the existing large butterfly valve and controls the column pressure using the small precision control valve. The VPC (valve position controller) controls the position of the large valve to keep the small valve in range. The VPC control loop is a self-regulating process where the hysteresis and deadband does not result in cycling. Tuned correctly, the new control strategy provides accurate pressure control using existing valves with no cycling at steady state.

Regulatory Control Optimization Results

Poorly implemented control strategies, faulty equipment, improper setup, bad installation, lack of anti-aliasing filters, and Murphy's Law are the conditions that require on-site system analysis testing to tune the regulatory control system.

Over the last several years, a major company's engineers have used the Protuner System Analyzer and the test procedures learned from the consultants to optimize unit operations at a number of facilities, yet they still find room for improvement. This clarifies the fundamental need for conscientious and consistent attention to the optimization of the regulatory control system to ensure economic APC ROI is achieved.

The tuning of the single and interactive loops is ascertained based on the analysis of the actual installed dynamic transfer functions of the individual processes. Tuning parameters are determined based on pole cancellation with adequate gain and phase margins, and damping factor requirements as a function of the installed linearity of each loop.

Feedforward tuning is based upon the actual dynamic transfer function models of the process and load disturbance.

The net result has been better, safe, and more efficient operating plants for this customer.

Advanced Process Control

Advanced control systems are implemented to control the process, not as individual levels, flows, pressures, or temperatures, but rather as each variable relates to productivity or efficiency of the process. From an advanced control system viewpoint, the control system is not single loop controls but a multi-variable envelope viewed as a polygon, with each side representing the constraints of pressure, temperature, *etc.* Within the envelope, the process is continuously maximizing efficiency.

There are a large number of techniques employed that come under the general category of advanced process control. The most common, yet least discussed advanced control strategy, is operator knowledge and confidence the regulatory control system works. In many cases, no matter what control strategy is implemented, operators will set the individual process variable setpoints at 'safe,' though not necessarily optimum target setpoints. Until operators are confident the regulatory control system is capable of safe and reliable operation-at or near process limits-the operator safe factor will often result in undesirable process integrity.

Another advanced-or optimization control strategy-is the proper application of regulatory control that is designed and implemented to maximize efficiency of the operation.

Examples include the use of variable speed pumps and correlated control strategies to allow the control system to better follow demand. Additional widely discussed advanced control strategies include dynamic matrix control, fuzzy logic, and multi-variable control. Each has several things in common including the obvious goal to continuously adjust regulatory control to maximize the operational efficiency of the process. Therefore, the success of an advanced control system is directly impacted by how well the regulatory control system functions.

Years of experience reinforce there is no magic panacea for the optimizing of a control system. In some cases, an advanced control package is sold as the ultimate answer to reducing variability and improving efficiency without adequate consideration given to the underlying regulatory control systems operation. Expectations of this nature are too frequently disappointing and can give process control an unjustified 'black eye.'

Most regulatory control systems are de-tuned at startup for steady state operation to mask design and equipment related problems, or because of a lack of knowledge about the processes dynamics. De-tuning the regulatory control system avoids 'troublesome oscillation,' but almost always results in the need to constantly adjust regulatory loop setpoints to overcome upsets 'caused' by the APC.

Management insight and commitment, trained and motivated people, the proper tools, and plain hard work on the regulatory portion of the control system form the necessary foundation to successfully apply APC.

The knowledge gained during testing and analyzing regulatory loops provides exactly the knowledge needed when developing, applying, and tuning the APC models and establish the foundation for continuous process improvements.

Chapter 8

A CONTROL AND MONITOR SYSTEM FOR THE GMRT

INTRODUCTION

Modern radio telescopes are complex assemblies of electronic and electro-mechanical subunits. To allow a successful observation, all of these sub-units have to be "set" as per the users requirements. For example, the antennas have to track the selected source, the front-ends have to be tuned to the chosen frequency band, all the amplifiers along the signal path have to be set at the value which would give the optimum signal to noise ratio, the local oscillators have to be tuned to select the desired frequency, the correlator has to be set up to do the appropriate fringe and delay tracking *etc*. In an interferometer like the GMRT this means that one has to, in a co-ordinated manner, control sub-systems which are several tens of kilometers separated from one another. In addition, it would be highly desirable for the health of the critical sub-systems to be able to be periodically monitored, so that should any subsystem fail, the affected data can be flagged, and also of course remedial action could be taken to fix the faulty unit. Further since it is not humanly possible to remember all the various safety limits of each of the sub-systems, one requires the telescope control system to not permit operations which could endanger the safety of the telescope or the human operators. The GMRT Monitor and control system was designed with all these different requirements in mind.

The GMRT control and monitor system allows one to :

1. Rotate of all the thirty antennas in azimuth and elevation, and/or to track a celestial source.

2. Bringing the required feed in the feed turret to the focus via the Feed Position System(FPS).

3. Select front end system parameters like the observing frequency band, desired noise calibration, *etc.*

4. Sets the IF sub-systems including the LO frequencies, the IF bandwidths and attenuation, the ALC operation *etc.*

5. Sets the baseband bandwidth and attenuation.

6. Monitor, literally, hundreds of system parameters at all points along the signal flow path.

7. Have a voice link between each antenna shell and the control room in the CEB (Central Electronics Building).

OVERVIEW

The major components of the Monitor and Control system are ONLINE (a unix level program that provides the user interface), PCROUTER, a PC based router, COMH, a communications handler that deals with the packet based communication between the Unix workstation on which ONLINE is running and the Antenna Based Computer (ABC, also called ANTCOM) located in each antenna shell, and finally several Monitor and Control Modules, (MCM) which provide the monitoring and control interfaces to the various sub-units (*i.e.* the LOs, amplifiers, *etc.*).

Fig. : Block diagram of the GMRT monitor and control System.

ONLINE

The ONLINE software running on a UNIX workstation provides the user interface for the control and monitor system. The commands typed by the user are sent to the relevant antenna(s) by the telemetry system. The monitoring data from all the various GMRT subsystems are also logged by ONLINE. Should some critical subsystem fail, ONLINE will raise an appropriate alarm so that remedial action can be taken.

PCROUTER

This converts the data from the format used for TCP/IP (ethernet) communication to one suitable for the serial communication links used by the GMRT telemetry system. As the name suggests this is PC based.

COMH

COMH is the communication handler and it handles all the communication between the UNIX workstation on which ONLINE is running and the various Antenna Base Computers (ABCs). COMH operates in a time division multiplexing (TDM) mode *i.e.* it sends the formatted user commands to the first antenna and then waits for an acknowledgment. If it receives an error free reply before the timeout period it selects the next antenna and the operation continues. In case COMH doesn't get a reply before the timeout period or if the reception is erroneous then it tries the same antenna again. After a total of three failures COMH passes on a Timeout or Checksum error (as appropriate) to ONLINE and then moves on to the next antenna.

ANTCOM or ABC

There is an ANTCOM (also called an ABC) located in each antenna shell. All communication between the antenna and ONLINE is routed through the ANTCOM in that antenna. The ANTCOM receives various parameters sent by COMH, performs some computations if necessary, and passes on the commands to the appropriate sub system of the antenna. The ANTCOM has three communication links, viz. (a) the main link between COMH & ANTCOM which operates at 250 kbps, (b) an asynchronous 9.6 kbps RS 422 communication link between ANTCOM & the Servo Control Computer (SCC) and (c) an asynchronous 9.6 kbps RS 485 communication link between ANTCOM & upto 16 Monitor and Control Modules (MCMs).

In addition to the ANTCOMs in the various antennas, there is also an ANTCOM (called ABC0) in the receiver room of the CEB. ABC0 handles the configuring of the baseband system in the receiver room.

SERVO CONTROL COMPUTER

In addition to an ANTCOM, each antenna has a Servo Control Computer (SCC) which is responsible for controlling the motion of the antenna. The SCC accepts movement commands, position information *etc.* from the ANTCOM, checks that the command is sensible, and if so obeys it. It also returns the antenna status information periodically through the same link. This information is passed on by the ABC to ONLINE and is displayed on a monitor in the CEB.

Monitor and Control Modules

MCMs are a general purpose Micro-controller based card which provides 16 TTL Control O/Ps and can monitor upto 64 analog signals. These MCMs are the interface to all the settable GMRT subsystems, like the front ends, the LOs the attenuators *etc.* At each antenna, MCM 5 is the interface to the front end system, while MCMs 2,3, and 9 are the interface to the LO and IF systems.

The Feed Positioning System (FPS) which is used to position the feed turret so that the desired feed is at focus is also controlled by the ANTCOM.

SIGNAL FLOW IN THE GMRT CONTROL & MONITOR SYSTEM

User commands for various antennas are processed by ONLINE running on a UNIX workstation and are sent to the COMH via the PCROUTER. The PCROUTER acts as a buffer and accepts the TCP/IP data on a 10/100 Mbps (*i.e.* a standard ethernet) link, strips the TCP/IP header and sends the data to COMH on a 38.4 kbps link. This uses a standard RS232-C link on the PC side and a conversion to RS 422 signals (differential TTL signals) on the COMH side.

COMH is basically an 80C186, 16 bit micro-controller based card, which works at a clock speed of 6 MHz. This card also contains a Zilog 85C30 dual channel communication controller. The two channels are respectively for SDLC/HDLC communication at 125 kbps (for communication with the ANTCOMs at the different antennas) and a asynchronous communication at 38.4 kbps (for communicating with the PCROUTER). COMH also has an Intel 29C17 CODEC (voice coder-decoder) to handle voice communication at 62.5 kbps, circuitry for digital Phase Lock Loop and other combinational logic to handle clock recovery and bit interleaving functions, as well as FSK modem chips NE 5080 and NE 5081 to handle FSK modulation and demodulation. COMH multiplexes command data, digitized voice, synchronization pulses, dial pulses and two aux channels into a single bit stream. This bit stream is then converted to an analog signal at 18 MHz.

Fig. : Block diagram of the GMRT communication handler (COMH). The antenna computer (ANTCOM) has essentially the configuration.

The block diagram for this multiplexing of voice and data. The structure of the multiplexed bit interleaved data frame. At the bottom of this figure is shown the flow diagram for the synchronous detector state machine.

Fig. : Block diagram of the voice and data multiplexer.

Fig. : Structure of the multiplexed voice and data frame. Shown at the bottom of the figure is the flow diagram for the synchronous detector state machine.

The FSK analog signal is sent via the fiber optic link to the ANTCOM at the antenna base. The ANTCOM has the same circuitry as COMH but unlike COMH it handles two serial communication links (using an INTEL 82510 Communication Controller) *i.e.* the ANTCOM-MCM communication link and a serial link to the Servo Control Computer (SCC). ANTCOM demodulates the FSK signal into 250 kbps data, regenerates the 250 kHz clock using a digital Phase Lock Loop, looks for sync bits and if it finds a match with no error or one bit error then it demultiplexes the data into command, voice, dial pulse and aux data and passes each to the appropriate circuit for further processing. The ANTCOM communicates with and controls the various subsystems in the antenna (other than the servo subsystem for which there is the dedicated SCC) via the MCM cards. A block diagram of the MCM card. The MCM card is a general purpose 80C535 Micro-controller based card which provides 16 TTL Control O/Ps and monitors upto 64 analog signals. It also has an RS485 communication link for communicating with the ANTCOM.

Fig. : Block diagram of the monitor and control (MCM) cards.

For the return link, the ANTCOM takes the monitoring information from SCC, MCMs and FPS forms a packet of SDLC/HDLC data and multiplexes with voice, hook status and aux channels into a single bit stream. This bit stream is converted into an FSK analog signal at 4.5 MHz, which is then up converted to 205.5 MHz using the regenerated 201 MHz as the LO. This analog signal is sent along with the astronomical signals to the CEB. At the CEB thirty CEBCOMs (one for each antenna) demodulate the FSK signal to convert it back into a digital 250 kbps data stream which is passed on to COMH via a 32 way multiplexer (MUX 32) card. The voice signals from the antennas are routed to the EPABX (telephone exchange) system. The block diagram for this telephonic communication. The voice signals are digitized using an INTEL 29C17 CODEC IC using a 7.8 kHz clock to produce a 62.5 kbps data stream. The CODEC uses "A law" for data companding/expanding.

Fig. : Block diagram of the telephony interfaces.

Error Detection

The error detection uses both Cyclic Redundancy Check (CRC) and checksum methods. SDLC/HDLC supports 16 bit CRC error detection. CRC can detect all the single errors, double errors and burst errors up to 16 bits in length and can also detect 99% of burst errors of lengths greater than 16 bits.

The way this works is as follows. A cyclic code message consists of a specific number of data bits $G(X)$ and a Block Check Character (BCC). Let equal the total number of bits in the message, equal the number of data bits, i.e. $n - k$ is the number of bits in the BCC. The code message is derived from two polynomials which are algebraic representations of two binary words, the generator polynomial $P(X)$ and the message polynomial $G(X)$. The generator polynomial $P(X)$ is a type of code used in CRC-12, CRC-16 and CRC-CCITT.

For example, n bits of binary data can be represented as a message polynomial of degree $n - 1$. Thus, an eight-bit long message 10101010 is represented as

$$G(X) = X^7 + X^5 + X^3.$$

The code message can be constructed as follows:

1. Multiply the message $G(X)$ by X^{n-k} where $n - k$ is the number of bits in the BCC.

2. Divide the resulting product $X^{n-k}[G(X)]$ by the generator polynomial $P(X)$.

3. Disregard the quotient and add the remainder $C(X)$ to the product to get the code message polynomial $F(X)$, which is represented as $X^{n-k}[G(X)] + C(X)$.

The division is performed in binary without carries or borrows. The code message $F(X)$ is transmitted as binary data and the receiver at the other end retrieves the message using the same generator polynomial and accepts the data if the remainder is zero.

SIGNAL MODULATION

The Control and Monitor system hardware essentially consists of a digital part, an analog part and the Optical Fiber system.

Fig. : Schematic of the GMRT telemetry system.

The optical fiber is a single mode analog link operating at 1310 nm, and can carry signals from a few MHz to about 1 GHz. There are two fibers (an 'forward link' and a 'return link') between the Central Electronics Building (CEB) and each antenna. In the forward link the telemetry signals use an 18 MHz carrier, and the return link has a 205.5 MHz carrier.

Fig. : Schematic of the signals carried by the forward and return link.

Frequency Shift Keying

The digital data that the telemetry system generates is converted to an analog signal using Frequency Shift Keying (FSK). FSK is a special type of modulation where the digital signals ("0" & "1") changes the frequency of the pseudo carrier to one of two frequencies, usually denoted as MARK and SPACE respectively.

If T is the duration of a bit, then the bandwidth (BW) occupied by the FSK signal is:

$$\left[v(\text{MARK}) + \frac{1}{T}\right] - \left[v(\text{SPACE}) - \frac{1}{T}\right] = v(\text{MARK}) - v(\text{SPACE}) + \frac{2}{T}.$$

For example, in the Forward Link, v (mark) = 19 MHz, v space) = 17 MHz and t = 4 microseconds (*i.e.* corresponding to a data rate of 250 kbps). Therefore, the bandwidth of the FSK signal in the forward link is

$$\Delta v = (19 - 17) + \frac{2}{4 \times 10^{-6}} = 2.5 \text{ MHz}.$$

AN OVERVIEW OF THE GMRT CORRELATOR

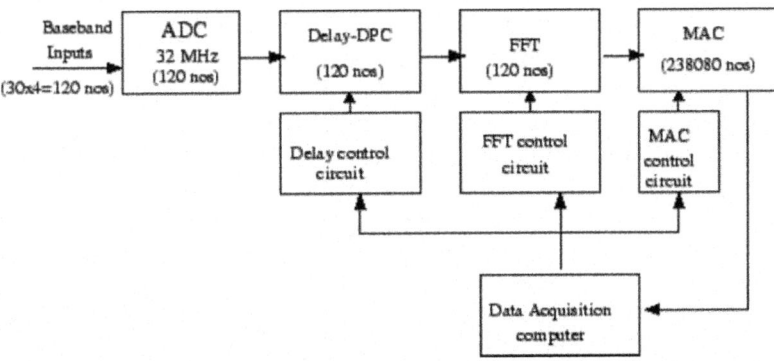

Fig. : A simplified block diagram of the GMRT correlator.

A simplified block diagram of the GMRT correlator. The basic units are the analog to digital converters (ADC), the Integral Delay compensation (Delay-DPC) subsystem, the Fourier transform and fractional delay compensation (FFT) subsystem and the multiplier-accumulator (MAC) unit. The data from the MAC output is acquired using a special purpose PC add on card. All of the subsystems, except the ADC, have DSP (digital signal processor) based control circuits. These control circuits are in turn controlled by the data acquisition computer.

ADC

The GMRT correlator uses 6 bit, uniform quantization ADCs. The ADCs are designed such that a Gaussian random signal of 0 dBm power will have minimum distortion and operate at a fixed clock frequency of 32 MHz. This means that when the input signal has a bandwidth of 16 MHz the digitized signal is Nyquist sampled. However at the GMRT, the input signal could have a bandwidth less than 16 MHz, for these signals the Delay-DPC effectively resamples the digitized signal so that down stream of the Delay-DPC unit the signal is Nyquist sampled.

Delay-DPC

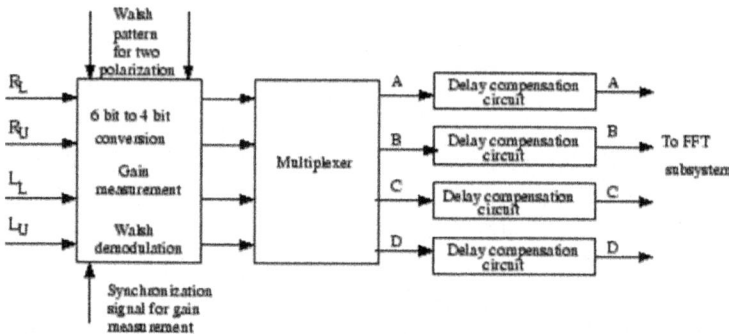

Fig. : Block diagram of the delay-DPC unit of the GMRT correlator.

The block diagram of the delay and data preparation card (delay-DPC). Each basic unit of the delay-DPC takes the four outputs of ADCs corresponding to the signals R_U, R_L, L_U and L_L from a given antenna. These 6 bit quantized signals are rounded off to 4 bits and then sent to a multiplexer. The multiplexer has various modes; for example any one of the four inputs of the multiplexer can be mapped to all four of its outputs. Other mappings include (a) $A = R_L$, $B = L_L$, $C = L_L$, $D = R_L$, and (b) $A = R_U$, $B = L_U$, $C = L_U$, $D = R_U$, which are used for polarization observations with the correlator. The multiplexer outputs are passed through a memory based integral delay compensation circuit. The delay compensated outputs are then fed to the FFT subsystem.

The rate at which data is written to the memory in the dly-DPC card is tunable. In particular it can be any one of $32/2^k$ MHz, where $k = 0$ to $k = 7$. This rate is chosen to be the Nyquist rate for the input signal bandwidth, *i.e.* for bandwidths smaller than 16 MHz, the rate is less than 32 MHz. However, the data is always read out at a constant rate of 32 MHz. To maintain the data throughput, data from the memory hence has to be read out in an 'overlapping' fashion. This way of reading the data provides the facility to perform 'overlapping' FFTs (and hence an improvement in the signal to noise ratio) when the input bandwidth is less than 16 MHz.

The two other functions of the delay-DPC system are (a) gain measurement (b) Walsh demodulation.

FFT

The block diagram of the FFT subsystem. The basic unit of the FFT subsystem takes two data streams from the Delay-DPC. In the first stage, a weighting function can be applied to the 4 bit time series. The weighting function is software selectable, and can be chosen to be one of the standward "window functions". This is followed by a number controlled oscillator (NCO), which does the fringe stopping. The two fringe stopped time series are passed through two sets of FFT engines, realized using VLBA ASICs, to perform Fourier transforms. Phase gradients are then applied to the spectrum of the signal to correct for delays smaller than the sampling interval (FSTC).

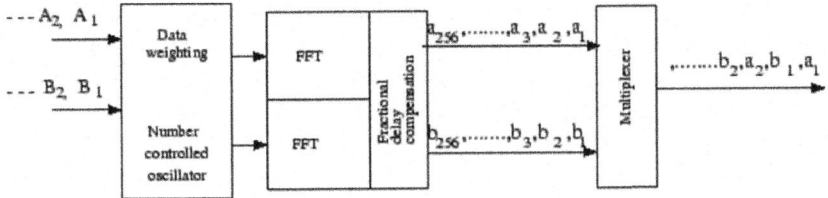

Fig. : Block diagram of the FFT unit of the GMRT correlator.

Each FFT engine can perform a Fourier transforms of maximal length 512 points. This length is software selectable to be 256, 128 or 16 points; it is even possible to bypass the FFT operation altogether. A 512 point FFT gives 256 channels, however in the next stage of the correlator (MAC) there are only enough multipliers for 128 channels per sideband per polarization. In the standard mode of operation, two adjacent FFT channels are hence averaged together in the MAC. A single MAC also acquires data from two FFT engines in a time multiplexed fashion. The data is multiplexed, where a_i and b_i are the spectral channels from the two FFT engines.

MAC

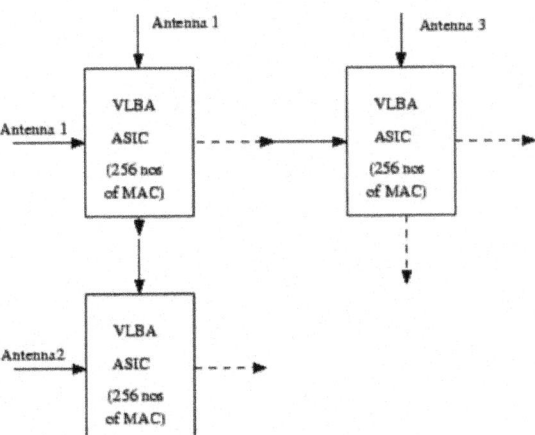

Fig. : Block diagram of the Multiplier-Accumulator (MAC) unit for the GMRT correlator.

The Multiplier and Accumulator (MAC) is, hardware wise, the most complex subsystem of the correlator. The MAC takes the FFT outputs computes the cross and self products and accumulates them for a maximum of 128 ms and a minimum of 4 ms. A schematic of the configuration of the multipliers. Each MAC unit consists of 256 accumulators. The MAC can be configured in several different modes. The flexibility allows the GMRT corrrelator to be used to make a wide variety of observations. Data from the MAC unit is read out by the Data Acquisition System (DAS) using a special purpose add on card on a PC.

MODES OF OPERATION OF THE GMRT CORRELATOR

The total number of multipliers available in the correlator is less than that required for the measurement of all four stokes parameters in all spectral channels for all sidebands of all antennas. Instead,the correlator is configurable in various ways. Some configurations would sacrifice polarization measurements for improved spectral resolution, while others allow the measurement of all four stokes parameters at the expense of total bandwidth. The most commonly used configurations of the correlator.

Non-Polar Mode

In this mode of observation the visibility for only one of the two polarizations can be measured in each of 256 spectral channels for all baselines (including self correlations). The maximum bandwidth possible is 2×16 MHz. Thus the observation will have half the total sensitivity of the GMRT.

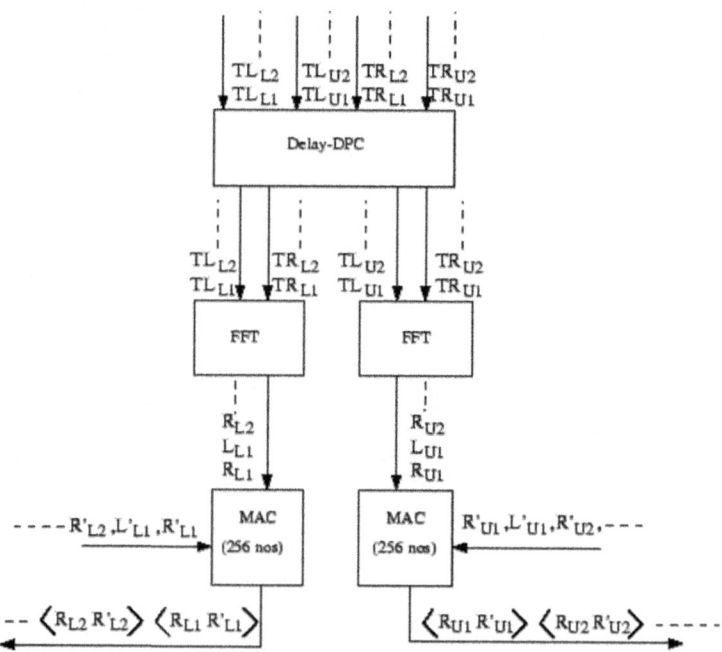

Fig. : The signal flow in the GMRT correlator for the Non-Polar mode. Signal names preceding with a T indicate time series and angular brackets denote time average. Numeric subscripts indicate the sample number for time series data and the frequency channel number for spectral data.

In the non-polar mode the required sampling frequency is selected in the delay-DPC system. The multiplexer in the delay-DPC is configured such that the data flow. TR_{ui}, TR_{Li}, TL_{ui} and TL_{Li} are the time series of the four baseband signals for a given antenna. The FFTs of these time series are R_{ui}, R_{Li}, L_{ui} and L_{Li} respectively (for $i = 1$ to 256). R'_{ui}, L'_{ui} are the corresponding signals from a second antenna. The MAC mode is selected such that the 256 channels in one of the sidebands of one of the polarizations (in this case R_{ui}) is integrated in its 256 accumulators. A second set of MACs integrate the signals from the second sideband of the same polarization (in this case R_{Li}).

Indian-Polar Mode

In this mode the visibility from both polarizations of all 30 antennas is measured but the number of spectral channels per baseline is limited to 128. Thus the spectral resolution is half that of the Non-Polar mode but the maximum bandwidth that can be observed in this mode is 32 MHz. Thus the observation will have the full sensitivity of the GMRT. For a non polarized source, this mode measures stokes I, and it is the most commonly used mode in interferometry.

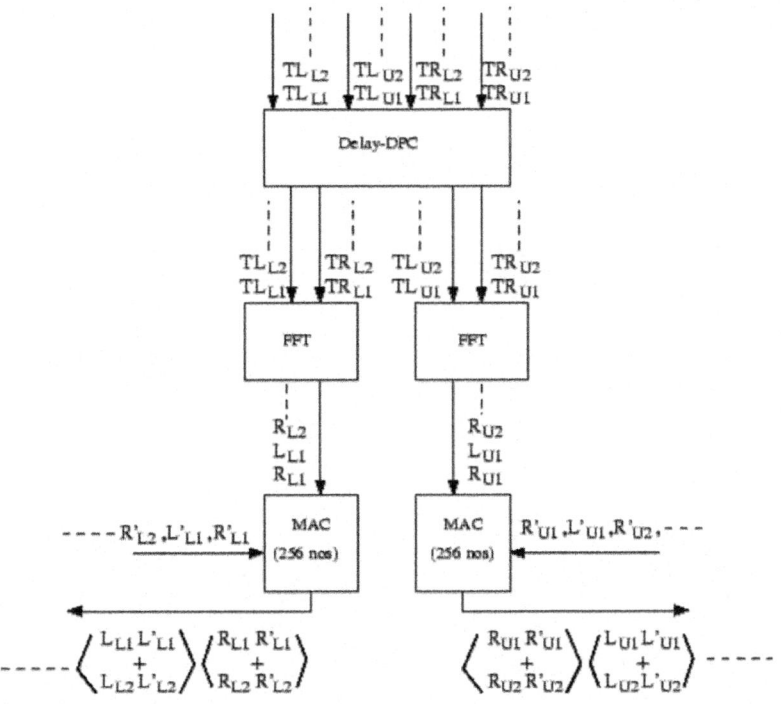

Fig. : Signal flow in the GMRT correlator when it is configured in Indian-Polar mode.

The Delay-DPC configuration for this mode is similar to that of the Non-Polar mode. The MAC is configured such that the adjacent channels of the same polarization are averaged together, thus reducing the number of channels from

256 to 128. Since each MAC unit has 256 accumulators, the other 128 accumulators of the same MAC are used to integrate the data from the second polarization of the same baseline.

Polar Mode

All the cross products needed for measuring the four Stokes parameters are measured in this mode. The number of channels available per baseline is again restricted to 128 and further one sideband from all 30 antennas is processed. Thus the maximum possible bandwidth in the Polar Mode is 16 MHz, as opposed to 32 MHz in the Indian Polar mode (which measures Stokes I for unpolarized sources), and the spectral resolution is also half of the maximum possible in the Indian Polar Mode.

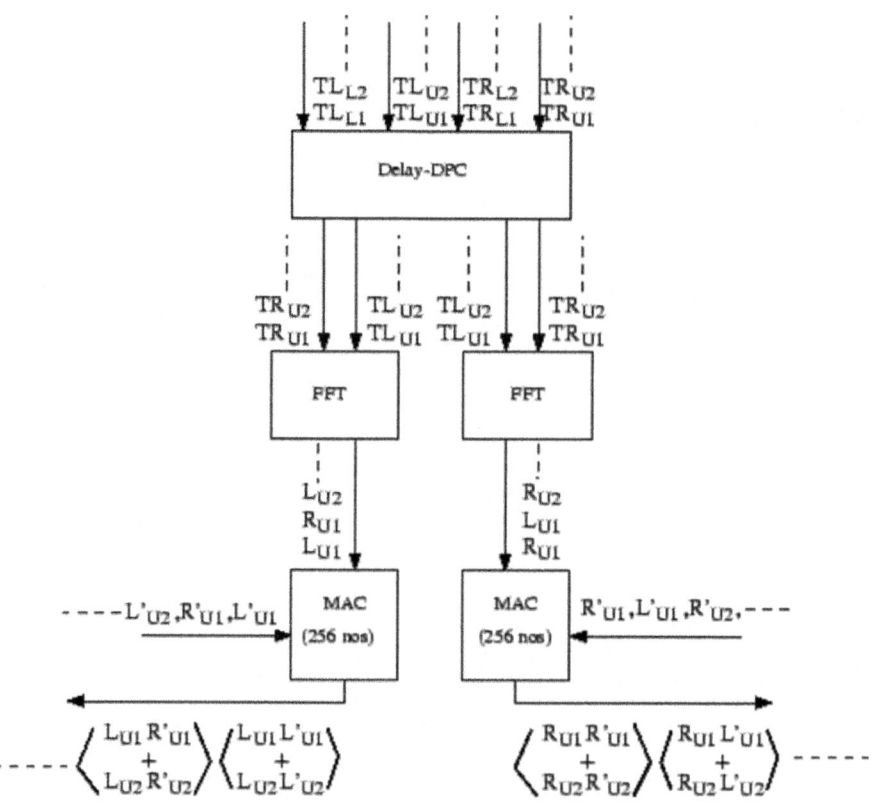

Fig. : Signal flow in the GMRT correlator in the Polar Mode. Signal names preceded by a T indicate time series and angular brackets denote time average.

The delay-DPC multiplexer is configured so that the data flow. The data from one side band for both polarizations (in this case R_{u}, L_{u}) is multiplexed to get the required data sequences. The MAC is configured in the polar mode such that it measures the cross product of the two polarizations in addition to the cross products of a polarization with itself. Adjacent channels of the cross product of one of the polarizations (*e.g.*: $R_{ui} \times R'_{ui}$) are averaged and integrated in 128 accumulators of the MAC. Unlike in the Indian-Polar mode, the second set of 128 accumulators integrate the cross product of the two polarizations (*e.g.*: $R_{ui} \times L'_{ui}$). Similar measurements of the second polarization (*i.e.* in this case, $L_{ui} \times L'_{ui}$ and $L_{ui} \times R'_{ui}$) are made in the second MAC. Thus all required cross products are measured.

THE DATA ACQUISITION SYSTEM FOR THE GMRT

The Giant Meterwave Radio Telescope (GMRT), an array of thirty antennas, situated at Khodad, Narayangaon, is by and large completely controlled by observers, sitting at their desks, with the help of a centralized networked system of computers at the Central Electronics Building (CEB).

The software system that achieves this can be broadly divided into two parts,

(i) one that deals with the control and monitoring of the antennas (ONLINE) and

(ii) one that deals with the control, monitoring as well as data acquisition (and processing) of the digital backends (DAS).

Both of these systems have a large number of small embedded subsystems that can be controlled across the network.

Any reasonable Data Acquisition System (DAS) for the GMRT must fulfill the following major requirements.

1. Interfacing with the hardware for acquisition of data.
2. Control of the correlator both for configuration as well as required periodic updates of parameters.
3. Monitoring of the status of the correlator and flagging the data appropriately.
4. Online processing of the data for reduction and archiving.

Data Acquisition

Let us first estimate the maximum data rate produced by the GMRT correlator. For each antenna there are four associated data streams (two sidebands USB and LSB, in each of two polarizations RR and LL) each of 16 MHz bandwidth. Therefore, one requires four sampling points per antenna with sampling rate of 32 MHz. The present correlator has only 60 sampling points, and handles only one side band. For each stream there is a corresponding FFT unit in the correlator. This FFT unit carries out a 512 data point transform in real time, *i.e.* one 512 point transform every 16μ sec. The 512 point transform corresponds to 256 complex numbers, *i.e.* 256 frequency channels. The time to acquire 512 data samples, *i.e.* 16μ sec is called a FFT cycle. The number of distinct pairs of antennas (including an antenna with itself, *i.e.* self correlations) that can be made from 30 antennas is $30 \times (30 + 1)/2 = 465$. Therefore, 465 Multiplier and Accumulator (MAC) units are required to correlate all data from 30 antennas. Each MAC unit accepts 4 data streams *i.e.* two polarizations from two antennas (it makes no sense to correlate USB with LSB) and multiplies them to produce 128 complex numbers each for two polarizations, or 256 values for one polarization. In either case, it is 256 complex numbers, which the MAC units sum for a duration of a STA (Short Term Accumulation) cycle, which is 4096 FFT cycles. One STA cycle is equivalent to $4096 \times 16\mu$ sec = 66 ms. The MAC data format is such that 4 bytes encode one complex number. The actual number of MACs in the correlator is 176×3 (there are 3 Racks with 176 MAC units each) or 528, *i.e.* there are $528 - 465$ redundant MACs. Therefore the total amount of data produced per second per side band is $528 \times (1\text{sec}/66 \text{ ms}) \times 256 \times 4$ bytes = 8 MB. The total data including both side bands would be = 16 MB/sec.

16 MB/s is a huge data rate to be sustained on any general purpose machine, or to be stored on any media. This means that would would need another piece of hardware in the correlator to carry out Long Term Accumulation (LTA). Such a hardware element was planned, but has not been implemented so far. Instead, in the present correlator system, the STA cycle has been configured for 8192 FFT cycles, *i.e.* the STA cycle duration is 132 ms. This brings down the sustained data rate per side band to 4 MB per second. Even this requires a special interface cards to input the data into the general purpose machine for processing. The host computer currently used for the DAS is a pentium based machine running the Linux operating system.

CORRELATOR CONTROL

The correlator system for the GMRT serves the following four major functions depicted.

Fig. : The schematic block diagram of the correlator. The four major blocks of the hardware as well as the host computer are shown.

1. Analog to Digital Conversion.

2. As we mentioned earlier that the full GMRT would require $30 \times 4 = 120$ sampling points, for thirty antennas with two side bands in each of two polarizations. The current correlator system suffices half of this requirement. The sampling takes place at 32 MHz in order to have data for a maximum of 16 MHz bandwidth. There is no control required for this unit of the correlator.

3. Delay DPC unit.

4. Signals originating from a given point in the source, and received via two different antennas, traverse different path lengths before they arrive at the samplers. This different path lengths arise because of the different locations of the two antennas as well as the different cable lengths between the two antennas and the correlator (and are called the geometric delay and the fixed delay respectively). The geometric delay changes with the Hour Angle of the source. In order to compensate for the differential delays that the signals from different streams have suffered, the Delay unit has to be periodically updated with the current values of time delays that have to be applied. Therefore, delay values are required to be transmitted from the host computer down to this unit of the correlator periodically. The Delay unit of the correlator can delay only for the integral number of sampling time intervals, which is 1/32MHz. Finer delay corrections are made in the FFT unit, where delay values are converted into a phase gradient across the band.

5. FFT unit.

   This chapter of the correlator carries out 512 points FFT every 1/32 MHz $\times$ 512 = 16μ sec. The two other major tasks that this unit performs are,

   o fringe phase subtraction, and

   o fractional sampling time delay correction.

   This requires periodic updates of the phase and FSTC values that are to be applied, which has also to be supplied by the host computer.

6. MAC unit.

7. The MAC unit can be configured in a variety of modes. This configuration is usually done by the host computer during the initialization sequence.

All these units (except the samplers) need to be initialized. The exact initialization required depends on the observing mode. To achieve this, at the start of the observation appropriate pieces of programs are loaded into the controllers, which then behave like embedded systems.

Monitoring the Health of the Correlator

The quality of the data acquired, depends on the health of the correlator for the duration of the observation. It is desirable to have flag bits in the recorded data indicating the state of the correlator at the time of the observation. There exists a planned set of parameters that are to be monitored and a method of transmitting such information to the host computer exists, but the actual monitoring has not been implemented yet. Hopefully, this will be done sometime in the future.

Chapter 9

TREATMENT OF ACTUAL CHEMICAL WASTEWATER BY A HETEROGENEOUS FENTON PROCESS USING NATURAL PYRITE

Liang Sun, Yan Li and Aimin Li*

State Key Laboratory of Pollution Control and Resources Reuse, School of the Environment, Nanjing University, Nanjing 210023, China; E-Mails: sunliangphd@tju.edu.cn (L.S.); liyan_0921@126.com (Y.L.)

* Author to whom correspondence should be addressed; E-Mail: liaimin@nju.edu.cn; Tel./Fax: +86-25-8968-0377.

Academic Editors: Rao Bhamidiammarri and Kiran Tota-Maharaj

ABSTRACT

Wastewater from chemical plants has remarkable antibiotic effects on the microorganisms in traditional biological treatment processes. An enhanced Fenton system catalyzed by natural pyrite was developed to degrade this kind of wastewater. Approximately 30% chemical oxygen demand (COD) was removed within 120 min when 50 mmol/L H_2O_2 and 10 g/L natural pyrite were used at initial pH from 1.8 to 7. A BOD_5/COD enhancement efficiency of 210% and an acute biotoxicity removal efficiency of 84% were achieved. The COD removal efficiency was less sensitive to initial pH than was the classic Fenton process. Excessive amounts of pyrite and H_2O_2 did not negatively affect the pyrite Fenton system. The amount of aniline generated indicated that nitrobenzene reduction by pyrite was promoted using a low initial concentration of H_2O_2 (<5 mmol/L). Fluorescence excitation emission matrix analyses illustrated that H_2O_2 facilitated the reduction by natural pyrite of organic molecules containing an electron-withdrawing group to electron-donating group. Thus, the Fenton-like process catalyzed by pyrite can remediate wastewater containing organic pollutants under mild reaction conditions and provide an alternative environmentally friendly method by which to reuse natural pyrite.

Keywords

Natural pyrite; heterogeneous Fenton process; chemical wastewater; reduction.

1. INTRODUCTION

Wastewater from chemical plants is a significant source of environmental contamination [1–3]. This kind of wastewater has remarkable antibiotic effects on microorganisms in traditional biological treatment processes because various organic compounds in the wastewater, especially nitro-aromatic compounds such as nitrobenzene, are toxic and bio-refractory [4,5]. Therefore, technologies that improve the biodegradability of this type of wastewater should be developed.

Various methods have been proposed to address this challenge. Physical-chemical technologies have been used to modify the chemical characteristics of wastewaters and render them treatable in biological systems without adverse effects. Advanced oxidation processes, such as Fenton reactions [6,7], ozonation [8,9], and photochemical oxidation [10,11] have shown great potential due to their high efficiency in removing refractory compounds. These technologies can mineralize contaminants completely and have been intensively investigated. In particular, the Fenton reactions have been proven to be one of the best choices for practical application because of their high efficiency, simple operation, and low cost. However, in spite of the high oxidation performance, the classic Fenton reaction (which is catalyzed by soluble Fe^{2+}) has some critical limitations: The operation needs to start under low initial pH (the optimum pH usually is 3) and the stoichiometric quantity of soluble Fe^{2+} that must be added generates significant amounts of sludge [12–14]. In the last few decades, the use of iron-bearing oxides (instead of soluble Fe^{2+}) as heterogeneous catalysts has received increasing attention as a means by which to overcome these drawbacks. Several studies have reported that the use of iron-bearing oxides as catalysts in Fenton reactions has advantages of low cost and easy operation, and may exhibit excellent catalysis performance in the removal of organic contaminants [15–17].

Pyrite (FeS_2) is the most abundant metal sulfide on the surface of Earth [18] and can be an appropriate material to act as a heterogeneous catalyst in the Fenton reaction. For example, Matta and Arienzo used a pyrite Fenton system for the oxidative degradation of 2,4,6-trinitrotoluene and reported that the observed degradation kinetics were much faster than those in the presence of other iron minerals such as magnetite and ferrihydrite [15,19]. Che and Bae used a pyrite Fenton system for the oxidative degradation of trichloroethylene and diclofenac, and reported that the degradation of both compounds was better in the pyrite

Fenton system than in a classic Fenton system [20,21]. Wu found that hydrogen peroxide (H_2O_2) enhanced by natural pyrite had great activity in the decoloration of azo dyes [22]. Zhang demonstrated that the degradation of nitrobenzene in the pyrite Fenton system was significantly enhanced compared to that achieved in a classic Fenton system [23].

However, there is little in the literature describing the application of the pyrite Fenton process to treat actual wastewater. The performance of this technology in remediating chemical wastewater is not yet known. Data are needed that define the catalytic performance of the pyrite Fenton process for the treatment of chemical wastewater so that the feasibility and application of this technology can be evaluated.

Meanwhile, reducing the quantity of contaminants may not necessarily be effective in reducing health and environmental risks because some degradation products may be more toxic than their parent compounds [24]. Several examples of this case have been reported in wastewater treatment processes [2,24]. Thus, information on the biodegradability and biotoxicity of chemical wastewater treated using the pyrite Fenton process is essential to evaluating the ecological safety and overall feasibility of this technology.

In the present study, the heterogeneous Fenton process using natural pyrite has been developed. The reactivity performance of the pyrite Fenton system was compared to that of a classic Fenton system when given the same initial conditions (iron content, H_2O_2 concentration, initial pH), and the effects of the dosage of pyrite and H_2O_2 on the removal of chemical oxygen demand (COD) were evaluated in detail. The biodegradability and biotoxicity of the treated wastewater were also assessed. Fluorescence excitation emission matrix (EEM) analysis was used to characterize the change of functional groups in the wastewater before and after treatment using the pyrite Fenton process. The results showed that adding a small quantity of H_2O_2 could enhance the reducing performance of the natural pyrite. The results contribute to a better understanding of the mechanism and reaction process of the pyrite Fenton technology. This report is the first to describe the reduction of nitrobenzene through Fenton oxidation catalyzed by natural pyrite.

2. EXPERIMENTAL SECTION

2.1. Reagents and Wastewater

Chemicals used in the experiments consisted of reagent grade (AR) H_2O_2 (30%), iron (II) sulfate heptahydrate ($FeSO_4 \cdot 7H_2O$), sulfuric acid (H_2SO_4, 98%), hydrochloric acid (HCl), and sodium hydroxide (NaOH) and were obtained from the Sinopharm Chemical Reagent Co. Ltd., Shanghai, China. All chemicals were used without further purification. The pyrite used in the experiments was mined from Anhui, China. The pyrite was sieved to a 300-mesh powder, washed with 1 mol HCl to remove surface oxidation layers, rinsed three times with deoxygenated deionized water and dehydrated with ethanol, and dried and stored in a closed vial under a pure nitrogen atmosphere. Elemental analyses showed that the iron content of the pyrite was 25%. The BET surface area of the pyrite was 5.960 m^2/g.

The wastewater samples used in experiments were collected from an industrial chemical plant located in the Jiangsu province in southeast China. The plant engages in the production of chemical intermediates for pharmaceuticals, dyes, and pesticides. The main products are P-nitrotoluene (PNT) and O-nitrotoluene (ONT). The wastewater samples were taken from the nitration process and nitrobenzene was the major by-product during the process. The characteristics of the wastewater are given in Table 1.

Table 1. Water quality indexes of wastewater used in experiments.

Index	COD (mg/L)	BOD$_5$/COD	TOC (mg/L)	Acute Biotoxicity (mg Zn^{2+}/L)	Nitrobenzene (mg/L)	pH	Conductivity (µS/cm)
Values	7500–8000	0.1	2000	471–490	>300	1.8	38,000

2.2. Degradation of Chemical Wastewater by the Pyrite Fenton System and the Classic Fenton System

Experiments on the pyrite Fenton and classic Fenton systems were conducted in 250-mL Erlenmeyer flasks. Chemical wastewater (200 mL) and a dosage of natural pyrite (2 g) were mixed, which was calculated to a solid-liquid ratio of 10 g/L, yielding an iron (as Fe(II)) concentration of 2500 mg/L (44.64 mmol/L). The degradation process was initiated by adding 1.02 mL of H$_2$O$_2$ (30%) into the flask, yielding an initial concentration of 50 mmol/L H$_2$O$_2$, after which the resulting slurry was mixed using a mechanical stirrer (150 rpm) at 25 °C. The classic Fenton system was operated under the same initial experimental conditions as for the pyrite Fenton system except 44.64 mmol/L of FeSO$_4$·7H$_2$O was used as an aqueous iron source instead of pyrite. Experiments were conducted at various initial pH (1.8–7), obtained by adding diluted sulphuric acid (10%) or sodium hydroxide solution (5 mol/L). Samples (1 mL) of solution were retrieved from each reaction flask at regular intervals (every 20 min) for further analysis using the techniques described below under "Analytical methods". Before the measurement of COD, BOD$_5$, acute biotoxicity, and the concentration of nitrobenzene and aniline, the pH of samples were adjusted to 8–9 to remove the residual aqueous solution iron and then they were aerated with N$_2$ for 30 min to remove the residual H$_2$O$_2$. The measurement of the concentration of Fe^{2+}, total aqueous Fe did not adjust the values of pH of samples. All experiments were performed in duplicate.

2.3. Effects of the Dosage of Pyrite and H$_2$O$_2$

The effects of the dosage of pyrite and H$_2$O$_2$ were investigated in batch experiments. All experiments were performed in Erlenmeyer flasks at 25 °C. To observe the effects of pyrite and H$_2$O$_2$ dosages, the pyrite concentrations were set as 5, 10, 20, and 50 g/L, and the H$_2$O$_2$ concentrations were set at 1, 2, 5, 10, 50, and 100 mmol/L. All experiments were performed in duplicate.

2.4. Analytical Methods

The COD concentration was measured by COD analyzer (HACH, DRB 200, Loveland, CO, USA), respectively [25]. The BOD5 was measured by the amount of oxygen consumed by the decomposition of organic matter in wastewater over 5 days [25]. The BOD5/COD index (B/C) was used to assess the wastewater biodegradability.

The acute biotoxicity measurements were conducted using the widely used photobacterium bioassay. This method quantifies the decrease in light emission of the bioluminescent bacteria *Photobacterium phosphoreum*, which results from exposure to pollutants. The extent of luminescence inhibition after 5 min exposure is standardized into an equivalent concentration of Zn^{2+}, which is used to express the degree of pollutant effects on the test bacteria. Thus, a greater luminescence inhibition corresponds to a higher equivalent Zn^{2+} concentration [26].

The concentration of Fe^{2+} was measured using to the 1,10-phenanthroline method [27]. The total aqueous Fe concentration was determined using flame atomic absorption spectrophotometry. The concentration of nitrobenzene was measured using the reduction azo-photometry [25], and the amount of aniline in the samples was determined using the N-(1-Naphthalene)-Ethylenediamine (N-(1-naphthalenyl)-1,2-ethanediamine; N-na; $C_{10}H_7NHCH_2CH_2NH_2$) method according to previous research [28].

All samples with no dilution were used for EEM analysis with Shimadzu UV-1800 ultraviolet-visible (UV/vis) spectrophotometer. The EEM fluorescence spectra were obtained as follows. The scanning field was set at an excitation wavelength from 245 to 400 nm and the emission wavelength from 280 to 500 nm, with 5- and 1-nm sampling intervals in excitation (Ex) and emission (Em) modes, respectively. All fluorescence data were presented in arbitrary units. A PARAFAC analysis was used to interpret the EEM data after first completing several preparative steps (*i.e.*, data loading, scattering removal, and initial explorative data analysis) [29–31].

3. RESULTS AND DISCUSSION

3.1. Comparison of COD Removal in the Pyrite Fenton System and the Classic Fenton System

As shown in Figure 1a, at initial wastewater pH of 1.8, approximately 30% of the COD was removed in the pyrite Fenton system within 120 min, whereas only 20% of the COD was removed in the classic Fenton system in the same reaction time. Figure 1b shows that at the optimal initial pH for operation of the classic Fenton process (pH 3 as reported in previous papers [6]), the COD removal efficiency was greater than 30% in both Fenton systems (in 120 min), but the classic Fenton system removed more COD than did the pyrite Fenton system. At an initial pH of 7, the classic Fenton system did not degrade the organics of the wastewater (Figure 1c); in contrast, the COD removal efficiency was still greater than 25% in the pyrite-Fenton system.

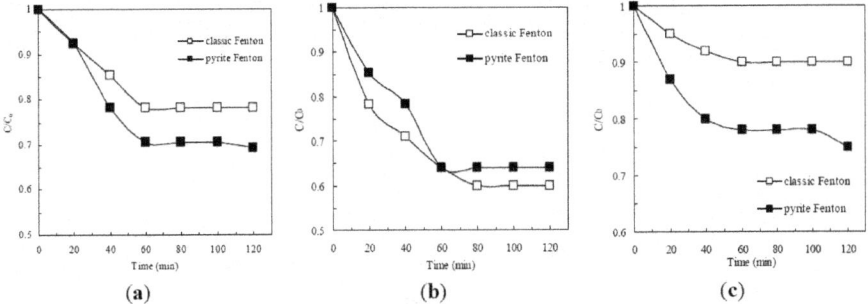

Figure 1. Comparison of pyrite Fenton and classic Fenton for the removal of COD under different initial pH. (a) initial pH = 1.8; (b) initial pH = 3; (c) initial pH = 7. Experimental conditions: (pyrite)0 = 10 g/L, $(H_2O_2)0$ = 50 mmol/L.

These results demonstrated that pyrite has a benefit as Fenton reaction media because the initial pH had little influence on the removal of organics in the pyrite Fenton process. Previous authors reported that two types of oxidation occur during the pyrite Fenton process: pyrite oxidation (Equations (1)–(3)) and Fenton oxidation (Equations (4) and (5)) [21–23].

$$2FeS_2 + 7O_2 + 2H_2O \rightarrow 2Fe^{2+} + 4SO_4^{2-} + 4H^+ \tag{1}$$

$$FeS_2 + 14Fe^{3+} + 8H_2O \rightarrow 15Fe^{2+} + 2SO_4^{2-} + 16H^+ \tag{2}$$

$$2FeS_2 + 15H_2O_2 \rightarrow 2Fe^{3+} + 4SO_4^{2-} + 2H^+ + 14H_2O \tag{3}$$

$$Fe^{2+} + H_2O_2 \rightarrow Fe^{3+} + HO\cdot + OH^- \tag{4}$$

$$Fe^{2+}OH^+ + H_2O_2 \rightarrow Fe^{3+}OH^{2+} + HO\cdot + OH^- \tag{5}$$

Mechanisms on the formation of H_2O_2 and $HO\cdot$ by pyrite in oxic aqueous solutions were investigated systematically in previous studies [21–23]. In the presence of oxygen, pyrite can react to produce Fe^{2+} in the solution (Equation (1)), then the aqueous Fe^{2+} reacts with H_2O_2 to form $HO\cdot$ and changes to Fe^{3+} (Equation (4)). Therefore, in the pyrite Fenton system, both the generation of aqueous Fe^{2+} (Equation (1)) and of $HO\cdot$ (Equation (4)) could be major rate-limiting reactions that affect the removal of wastewater COD. However, in the classic Fenton system, the generation of $HO\cdot$ (Equation (4)) is the only rate-limiting reaction. As shown in Figure 2, the aqueous total Fe concentration in the pyrite system gradually increased, which indicated that the aqueous Fe^{2+} was generated gradually. When the process was applied at the initial wastewater pH of 1.8, the aqueous total Fe concentration was 725 mg/L after 120 min treatment. In contrast, the aqueous total Fe concentration was only 195 mg/L at an initial wastewater pH of 7, which resulted in less iron sludge being produced. In addition, the less iron concentration could be reduced to 21.2 mg/L after the subsequent coagulation and sedimentation process.

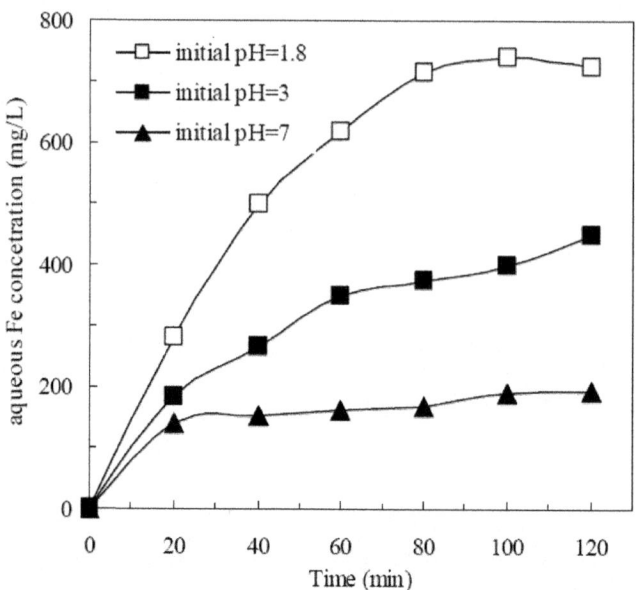

Figure 2. Aqueous Fe concentrations during the degradation in pyrite Fenton systems under different initial pH. Experimental conditions: $(pyrite)0 = 10$ g/L, and $(H_2O_2)0 = 50$ mmol/L.

According to Equation (4), the initial pH plays a key role in the removal of organics in the classic Fenton system. To avoid the precipitation of $Fe(OH)_3$, which consumes Fe^{2+} in the aqueous solution, the classic Fenton process is always operated at pH 3 (or lower) [32,33]. The present study demonstrated the validity of this conclusion. However, in the pyrite Fenton system, the COD removal efficiencies achieved at all initial pH values were quite similar, although the efficiency at pH 3 was somewhat superior to that at the other pH values. This phenomenon can be explained by the following reasons.

As a heterogeneous material, pyrite has the primary advantage of helping to avoid the formation of iron oxide sludge [34]. Thus, pyrite can expand the effective pH range in which the Fenton process can operate successfully [21,23]. The variation of suspension pH with respect to time in the pyrite Fenton system is shown in Figure 3. As the reaction proceeded, the pH of the suspension underwent obvious changes. When the initial wastewater pH was extremely acidic, the pH slightly (within a few minutes) increased as the reaction progressed, then remained in the acidic range. In contrast, when the initial wastewater pH was at neutrality, the pH first decreased drastically from 7 to 5.5, and then gradually decreased to pH 4.8. As described by Equation (1), the pyrite can react with oxygen and generate protons, which is the reason that the pH decreased. Importantly, the decreased pH helped sustain the Fenton reaction. Therefore, in contrast to the classic Fenton system, the pyrite Fenton system can naturally achieve the appropriate pH conditions for effective Fenton reactions without the need for chemical additives.

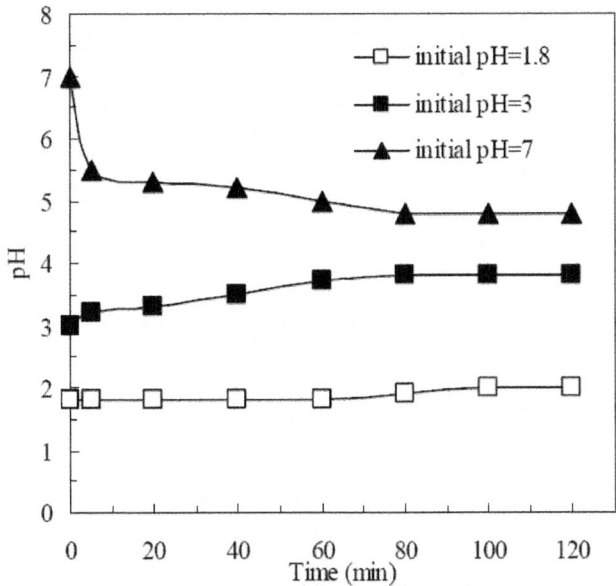

Figure 3. Variation of suspension pH with respect to time. Experimental conditions: (pyrite)0 = 10 g/L, and $(H_2O_2)0 = 50$ mmol/L.

3.2. Biodegradability Enhancement and Acute Biotoxicity Reduction by the Pyrite Fenton System

To investigate the effect of biodegradability enhancement and acute biotoxicity reduction via different Fenton system, batch experiments using pyrite, zero valent iron, and magnetite were set up. The changes in the biodegradability of wastewater as a result of treatment are illustrated in Figure 4. The B/C index of the untreated chemical wastewater was nearly 0.1, which demonstrated that this kind of wastewater was resistant to treatment using traditional biological technologies. Neither oxidation by H_2O_2 or reduction by natural pyrite as individual processes improved the biodegradability of the wastewater remarkably. The B/C index following H_2O_2 oxidation was 0.11, and was 0.13 following pyrite reduction. However, after wastewater was treated using the Fenton processes, the B/C index increased to approximately 0.3 in the classic Fenton system, to 0.31 in the pyrite Fenton system, and to 0.34 in the ZVI Fenton system, while it was only 0.23 in the magnetite Fenton system. Thus, the pyrite Fenton system showed a remarkable ability to improve wastewater biodegradability.

As shown in Figure 4, the COD efficiency of oxidation by H_2O_2 was 10% within 120 min, however, the value of BOD was decreased, suggesting that the biodegradability enhancement was mainly caused by COD removal. In the pyrite system, the COD was removed less, but the BOD was increased remarkably. The result showed that pyrite has an extremely strong reducing capacity for pollutant removal. The electron-deficient groups of compounds that include nitrobenzene could be reduced to electron-donating groups (such as aniline) to increase the biodegradability of wastewater. Comparing the results of COD removal and biodegradability enhancement of pyrite Fenton system, the biodegradability data demonstrated more remarkable changes than COD removal. These results indicated that the organic compounds were not degraded completely but were transformed to other less toxic compounds. During the Fenton process, the macromolecular compounds were split to small molecule substances by $HO\cdot$, and these reactions tend to simplify the structure of molecules to improve the biodegradability.

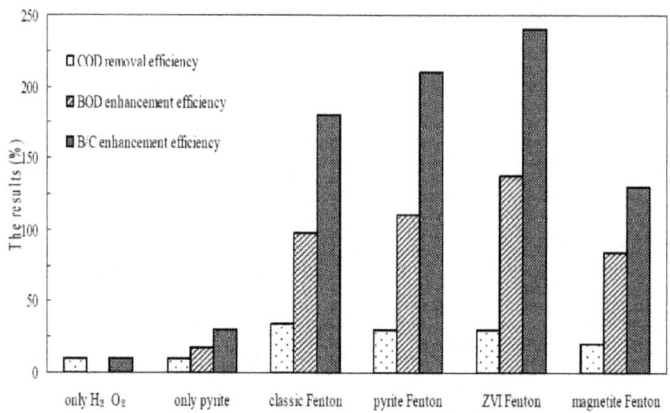

Figure 4. Comparison of pyrite Fenton and classic Fenton for the enhancement of B/C. Experimental conditions: $(pyrite)_0$ = 10 g/L, $(ZVI)_0$ = 10 g/L, (magnetite)0 = 10 g/L, $(H_2O_2)_0$ = 50 mmol/L, and intial pH = 3.

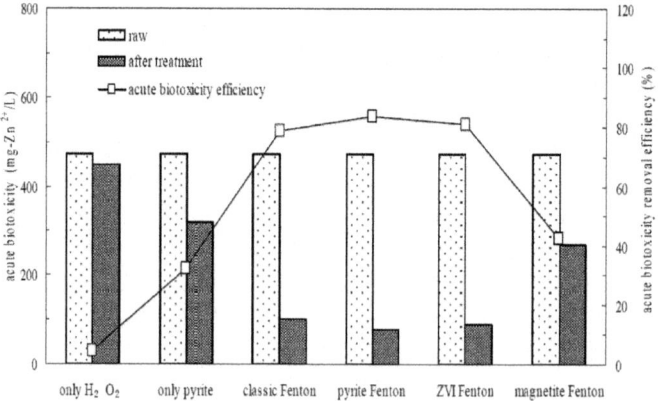

Figure 5. Comparison of pyrite Fenton and classic Fenton for the removal of acute biotoxicity. Experimental conditions: $(pyrite)_0$ = 10 g/L, $(ZVI)_0$ = 10 g/L, (magnetite)0 = 10 g/L, $(H_2O_2)_0$ = 50 mmol/L, and intial pH = 3.

Batch experiments further confirmed the biotoxicity reduction effected by wastewater treatment. The acute biotoxicity values of the wastewater before and after treatment are shown in Figure 5. The initial overall toxicity of the influent wastewater was 471.4 mg Zn^{2+}/L, which indicated that the wastewater posed a high environmental risk. As with their effects on wastewater biodegradability, neither H_2O_2 oxidation nor pyrite reduction acting alone reduced the acute biotoxicity of wastewater observably. The acute biotoxicity values of samples subjected to H_2O_2 oxidation and pyrite reduction were 450 and 320 mg Zn^{2+}/L, respectively. However, the acute biotoxicity values of samples were remarkably reduced to 100, 77 and 90 mg Zn^{2+}/L after the classic Fenton, pyrite Fenton and ZVI Fenton treatments, respectively. While it was 270 mg Zn^{2+}/L in the magnetite Fenton system.

The results indicated that the magnetite Fenton system can not significantly improve the biodegradability and reduce the biotoxicity of chemical wastewater, while both the pyrite Fenton and ZVI Fenton technologies are the novel alternative to the classic Fenton process for simultaneously improving the biodegradability and reducing the biotoxicity.

3.3. Effects of the Pyrite and H_2O_2 on the Removal of Organics in the Pyrite Fenton System

The effect of natural pyrite dosage on the removal of pollutants was examined in batch experiments using different initial dosages of pyrite (0–50 g/L). The results are shown in Figure 6. Within 120 min, the COD removal efficiency reached 10%, 20%, 36%, 32%, and 26% at pyrite dosages of 0, 5, 10, 20, and 50 g/L, respectively. These results indicated that COD removal increased significantly as the initial pyrite concentration increased from 0 g/L to 10 g/L, whereas COD removal decreased slightly as the initial pyrite concentrations increased from 20 g/L to 50 g /L. Based on these results, a pyrite dosage of 10 g/L was optimal for the pyrite Fenton system treating the particular wastewater used in the experiments. This variation in COD removal as a function of pyrite dosage probably can be attributed to the follow reason.

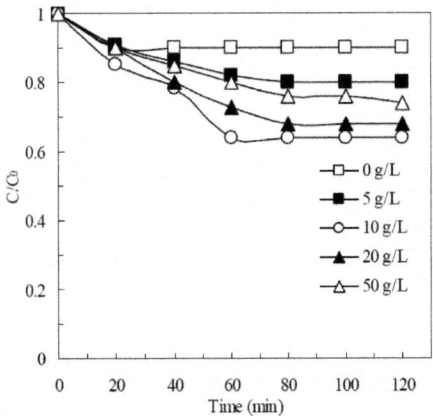

Figure 6. The effect of pyrite dosage on the removal of COD in pyrite Fenton system. Experimental conditions: $(H_2O_2)_0$ = 50 mmol/L, and initial pH = 3.

According to Equations (1)–(3), increasing the amount of pyrite in suspension proportionally increases the aqueous Fe^{2+} concentration, which improves $HO\cdot$ formation (Equation (4)) and enhances the degradation of organics. However, previous authors pointed out that an excessive amount of aqueous Fe^{2+} in a pyrite suspension may promote the unwanted consumption of $HO\cdot$ (as shown in Equation (6)), which can negatively affect the oxidative degradation of COD by Fenton reactions [21,22,32].

$$Fe^{2+} + HO\cdot \rightarrow Fe^{3+} + OH^- \tag{6}$$

Figure 7 shows the effect of the initial H_2O_2 concentrations on the removal of COD in the pyrite Fenton system. When the initial H_2O_2 was 1 mmol/L, the removal efficiency of COD was 15% after 120 min, but at an initial H_2O_2 concentration of 50 mmol/L, this efficiency increased to 36%. However, the TOC removal efficiency was only 10% under this condition. The results indicated that the organic compounds contained in the wastewater were not degraded completely but were split to small molecule substances by $HO\cdot$. The stoichiometric amount of H_2O_2 for the complete mineralization of the carbon content to CO_2 and H_2O following Equation (7) was calculated to 33.33 mmol/L. It is clear that the actual dosage of H_2O_2 is more than the calculated value. Segura $et\ al.$ also found this phenomenon in zero valent iron Fenton system [35]. They illustrated that the increase of the H2O2 showed a clear enhancement of performance when the highest oxidant loading (200% H_2O_2) was used. However, when the initial H_2O_2 increased to 100 mmol/L, the COD remove efficiency decreased to 30.

$$C + 2H_2O_2 \rightarrow CO_2 + 2H_2O \tag{7}$$

These variations were consistent with those observed in previous studies [6,36], which proposed that when an appropriate concentration of Fe^{2+} was provided in the reaction system, an increase of the H_2O_2 concentration could enhance the oxidative degradation of organics in the pyrite Fenton system due to the improvement in $HO\cdot$ formation. In contrast, an excessive amount of H_2O_2 in a pyrite suspension readily reacts with generated $HO\cdot$ (Equation (8)), reducing the oxidative removal of COD by $HO\cdot$.

$$H_2O_2 + HO\cdot \rightarrow OOH\cdot + H_2O \tag{8}$$

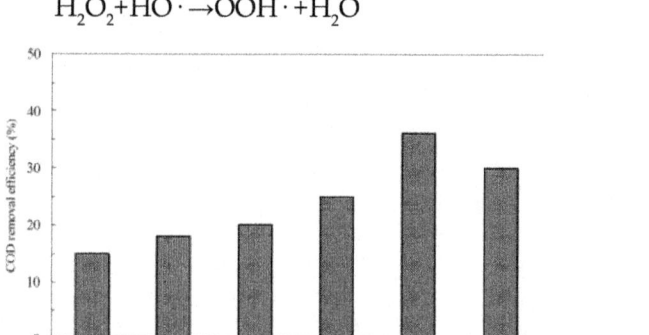

Figure 7. The effect of H_2O_2 concentrations on the removal of COD in pyrite Fenton system. Experimental conditions: $(pyrite)_0$ = 10 g/L, and initial pH = 3.

However, neither of the potentially negative effects posed by excessive amounts of pyrite and H_2O_2 appeared to be significant in the pyrite Fenton system. Relative insensitivity to such effects was another advantage that the pyrite Fenton system offered over the classic Fenton system.

3.4. The Generation of Aniline in the Pyrite Fenton System

Figure 8 shows the influence of initial H_2O_2 concentration on the generation of aniline. The untreated wastewater did not contain aniline, and no aniline was generated in the classic Fenton process. In the pyrite Fenton system, when only pyrite was added, the concentration of aniline in the treated wastewater was 18.7 mg/L after 120 min, which indicated that the natural pyrite could have a reduction capacity for nitro-aromatic compounds. In response to H_2O_2, within 120 min the concentration of aniline reached 23.4, 37.5, 9.4, and 3.7 mg/L at initial H_2O_2 concentrations of 1, 2, 5, and 10 mmol/L, respectively. However, when the initial H_2O_2 concentration was 50 mmol/L, no aniline was detected in the aqueous solution. These results showed that low initial concentrations of H_2O_2 (from 1 mmol/L to 2 mmol/L) significantly improved the reduction performance of pyrite, whereas high concentrations of H_2O_2 (5 mmol/L to 50 mmol/L) inhibited the reduction performance.

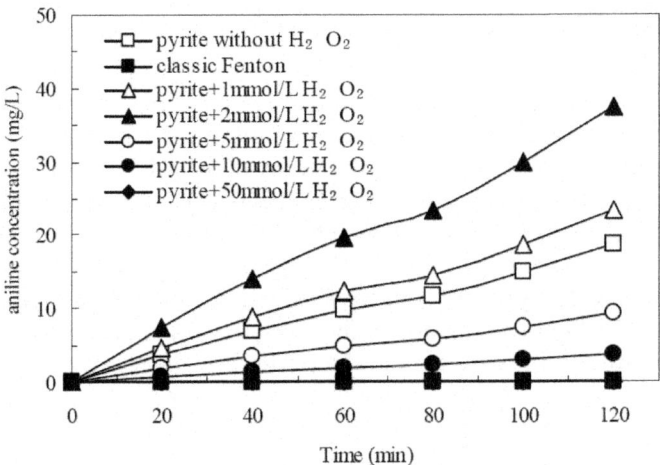

Figure 8. Amount of aniline in the aqueous solution with different initial H_2O_2 concentration. Experimental conditions: $(\text{pyrite})_0 = 10$ g/L, and initial pH = 1.8.

Stumm and James proposed that the dissociation of pyrite releases dissolved ferrous and sulfur ($S2^{2-}$) ions, which have an extremely strong reducing capacity for pollutant removal (Equation (9)) [37]. Zhang also found that the released dissolved ferrous effectively made pyrite an electron donor that resulted in the formation of Fe^{3+}, whereas the nitro group of compounds that include nitrobenzene were electron-deficient and thus, they could be readily reduced [34].

$$FeS_2 \rightarrow Fe_{(aq)}^{2+} + S_{2(aq)}^{-2} \tag{9}$$

Many studies have suggested that the decomposition of H_2O_2 can be controlled by surface-catalyzed process in the heterogeneous reaction involving iron oxides (*e.g.*, magnetite, goethite, and hematite) [38,39]. Kong pointed out that the oxygen and H_2O_2 in an aqueous solution could oxidize the surface Fe(II) of pyrite to Fe(III), as described in Equations (10) and (11) [40]. Kwan found that most organics in the heterogeneous Fenton process were degraded by $HO\cdot$ produced from dissolved Fe^{2+} (Equation (4)), not by a surface-catalyzed process (Equation (10)) [41]. Bae also observed this phenomenon in the pyrite Fenton system and furthermore demonstrated that H_2O_2 could reduce the surface Fe(III) of pyrite to Fe(II), and that the regenerated Fe(II) could be released into the aqueous solution and react with H_2O_2 to produce $HO\cdot$ (Equation (4)), which could react with target contaminants [21].

$$Fe_{pyrite}^{II}+O_2 \rightarrow Fe_{pyrite}^{III}+O_2^{-} \tag{10}$$

$$Fe_{pyrite}^{II}+H_2O_2 \rightarrow Fe_{pyrite}^{III}+HO\cdot+OH^{-} \tag{11}$$

$$Fe_{pyrite}^{III}+H_2O_2 \rightarrow Fe_{pyrite}^{II}+OOH\cdot+H^{+} \tag{12}$$

The pathway of this reaction process is shown in Figure 9. According to Equation (12), H_2O_2 can reduce the surface Fe(III) of pyrite to Fe(II) and enhance the generation of aqueous Fe^{2+}. When the initial concentration of H_2O_2 is low (such as 1 mmol/L and 2 mmol/L), the residual H_2O_2 in the aqueous solution cannot oxidize the aqueous Fe^{2+} thoroughly; hence, there is ample aqueous Fe^{2+} to join in the reduction of nitrobenzene compounds to aniline. In the present study, aqueous Fe^{2+} concentrations reached 198 and 187 mg/L in the pyrite Fenton system (after 120 min) at initial H_2O_2 concentrations of 1 and 2 mmol/L, respectively. However, when the initial H_2O_2 concentration was sufficient (greater than 10 mmol/L), the aqueous Fe^{2+} was oxidized by the excess H_2O_2, indicating that Fenton oxidization is the main reaction that occurs during the entire process. In fact, the mechanism and entire process of the pyrite Fenton technology is complex and not very explicit. Further investigation of the reductive transformations and oxidative pollutant degradation that occurs in the pyrite Fenton system is ongoing.

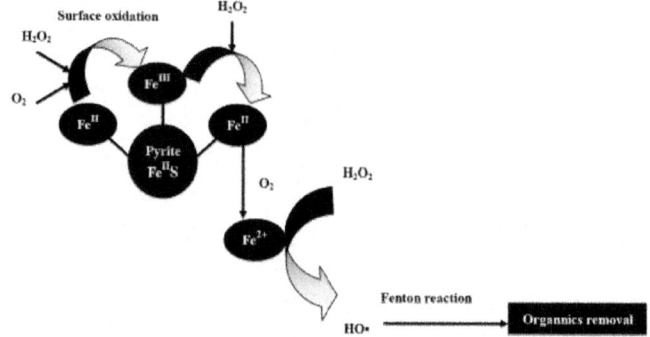

Figure 9. Proposed pathway for oxidative degradation of organics by pyrite Fenton system.

3.5. Evaluation of Organic Functional Groups by EEM in the Pyrite Fenton System

EEM has been frequently used to characterize organics in water and wastewater treatment systems due to its high sensitivity, good selectivity, and nondestructive effect on samples [42,43]. As a technique of multivariate data analysis, PARAFAC can mathematically decompose the complex fluorescence spectra into

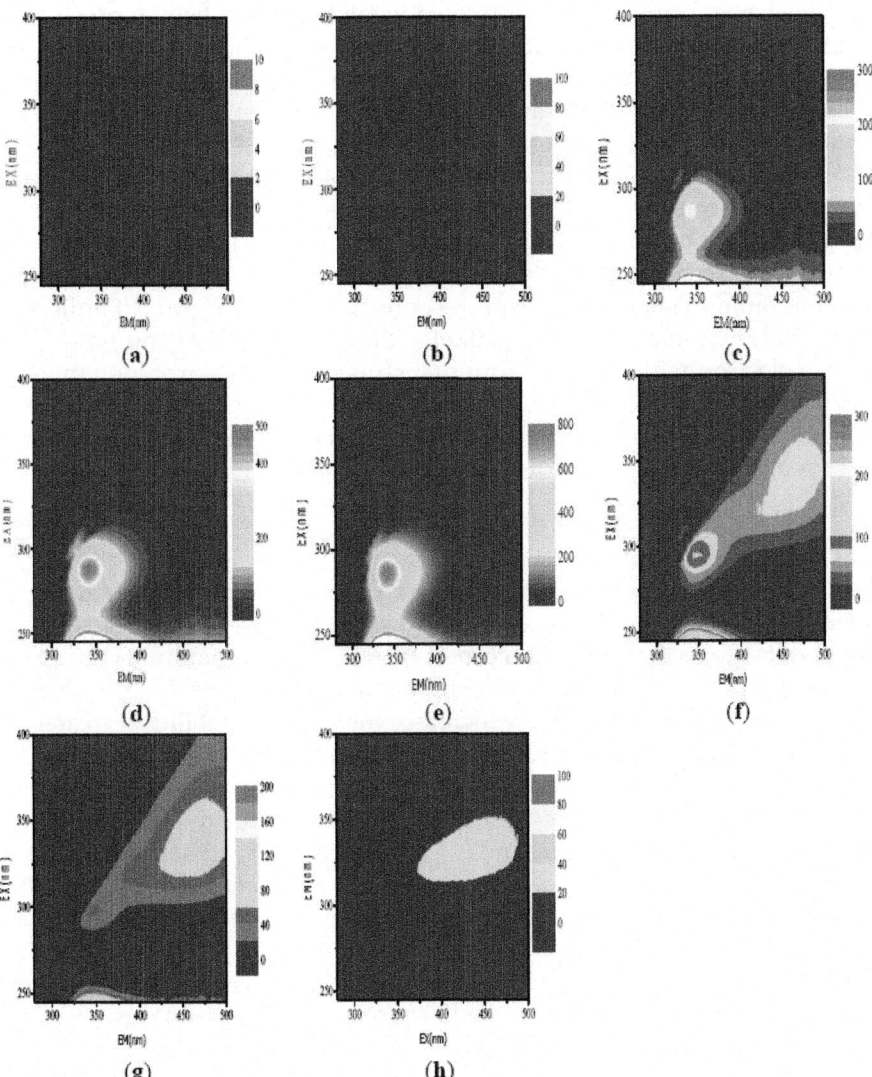

Figure 10. Sample results scanned by three-dimensional fluorescence. (a) raw water; (b) classic Fenton; (c) pyrite without H_2O_2; (d) pyrite + 1 mmol/L H_2O_2; (e) pyrite + 2 mmol/L H_2O_2; (f) pyrite + 5 mmol/L H_2O_2; (g) pyrite + 10 mmol/L H_2O_2; (h) pyrite + 50 mmol/L H_2O_2.

individual fluorescent components for both quantitative and qualitative analysis [44,45]. In the present study, the EEM-PARAFAC analysis method was used to evaluate the organic functional groups in the wastewater before and after it was subjected to classic Fenton and pyrite Fenton treatment.

Figure 10 shows the three-dimensional fluorescence spectra for various samples. Using the peak-picking function of the instrument's FL solution software, no peaks were found in the EEM spectra from either the raw (*i.e.*, untreated) wastewater or from wastewater treated using the classic Fenton process (Figure 10a,b); this indicated that there were no electron-donating groups contained in these samples. One major peak was found in the spectra from a number of samples at approximately Ex 290 nm and Em 432 nm (Figure 10c–g). The relative heights of the peaks for each sample were calculated to 211.1, 471.9, 772.2, 102.3, and 41.87, respectively. No spectral peak was observed for the sample that included pyrite and 50 mmol/L H_2O_2 (Figure 10h). Generally, fluorescent peaks at Ex 290 nm and Em 432 nm originate from amine groups ($-NH_2$), and the height of a peak corresponds with the quantity of organic material that contains amine groups (such as aniline) [46]. The spectral responses shown in Figure 10 indicated that low initial H_2O_2 concentrations facilitated the reduction of organic molecules containing electron-withdrawing groups (such as $-NO_2$) to electron-donating groups (such as $-NH_2$) by pyrite. These results proposed that increased H_2O_2 concentration could enhance the reduction of nitrobenzene in a pyrite Fenton system by improving the surface Fe(II) regeneration.

4. CONCLUSIONS

In this research, the treatment of actual chemical wastewater using an enhanced Fenton system catalyzed by natural pyrite was evaluated systematically. The results shows that the Fenton process catalyzed using natural pyrite effectively removes COD across a broad range of initial pH (pH 1.8 to 7). Furthermore, the pyrite Fenton process effectively increases the biodegradability of treated wastewater and simultaneously reduces the toxicity of treated wastewater. Compared to that of the classic Fenton process, the COD removal efficiency of the pyrite Fenton process is less sensitive to the initial solution pH and relatively insensitive to excessive amounts of both pyrite and H_2O_2. Finally, in the pyrite Fenton system, the reduction of nitrobenzene and organic molecules containing electron-withdrawing groups to electron-donating groups by natural pyrite can be promoted by adding a small amount of H_2O_2 (initial concentration <5 mmol/L). Thus, the pyrite Fenton process offers several operational advantages over the classic Fenton process and is a viable method by which to remediate organic pollutants in wastewater, while providing an environmentally friendly alternative for reusing natural pyrite.

Acknowledgments

We gratefully acknowledge the generous support provided by the Nation Water Pollution Control and Management of major special science and technology of China (No. 2015ZX07204002) and the National Natural Science Foundation of China (NO. 51408295), China.

Author Contributions

All authors were involved in discussing the ideas and designing this study. Sun Liang drafted the manuscript. Li Yan and Li Aimin edited draft versions and finalized the manuscript. All the authors have read and approved the final manuscript.

Conflicts of Interest

The authors declare no conflict of interest.

REFERENCES

1. Fan, J.H.; Ma, L.M. The pretreatment by the Fe-Cu process for enhancing biological degradability of the mixed wastewater. *J. Hazard. Mater.* **2009**, *164*, 1392–1397.

2. Sun, L.; Wang, C.; Ji, M.; Kong, X. Treatment of mixed chemical wastewater and the agglomeration mechanism via an internal electrolysis filter. *Chem. Eng. J.* **2013**, *215–216*, 50–56.

3. Wu, C.Y.; Zhou, Y.X.; Wang, P.C.; Guo, S.J. Improving hydrolysis acidification by limited aeration in the pretreatment of petrochemical wastewater. *Bioresour. Technol.* **2015**, *194*, 256–262.

4. Ren, C.X.; Li, Y.M.; Li, J.F.; Sheng, G.D.; Hu, L.J.; Zheng, X.M. Immobilization of nanoscale zero valent iron on organobentonite for accelerated reduction of nitrobenzene. *J. Chem. Technol. Biotechnol.*

5. Yin, W.Z.; Wu, J.H.; Huang, W.L.; Wei, C.H. Enhanced nitrobenzene removal and column longevity by coupled abiotic and biotic processes in zero-valent iron column. *Chem. Eng. J.* **2015**, *259*, 417–423.

6. Chamarro, E.; Marco, A.; Esplugas, E. Use of Fenton reagent to improve organic chemical biodegradability. *Water Res.* **2001**, *35*, 1047–1051.

7. Canizares, P.; Paz, R.; Saez, C.; Rodrigo, M.A. Costs of the electrochemical oxidation of wastewaters: A comparison with ozonation and Fenton oxidation processes. *J. Environ. Manag.* **2009**, *90*, 410–420.

8. Dariush, S.; Moheb A.; Abbas R.; Larouk S.; Roy R.; Azzouz A. Total mineralization of sulfamethoxazole and aromatic pollutants through Fe(2+)-montmorillonite catalyzed ozonation. *J. Hazard. Mater.* **2015**, *298*, 338–350.

9. Nakai, S.; Okuda, T.; Okada, M. Production of mono-and di-carboxylated polyethylene glycols as a factor obstacle to the successful ozonation-assisted biodegradation of ethoxylated compounds. *Chemosphere* **2015**, *136*, 153–159.

10. Kaur, M.; Verma, A.; Rajput, H. Potential use of foundry sand as heterogeneous catalyst in solar photo-fenton degradation of herbicide isoproturon. *Int. J. Environ. Res.* **2015**, *9*, 85–92.

11. Doumic, L.I.; Soares, P.A.; Ayude, M.A.; Cassanello, M.; Boaventura, R.A.R.; Vilar, V.J.P. Enhancement of a solar photo-Fenton reaction by using ferrioxalate complexes for the treatment of a synthetic cotton-textile dyeing wastewater. *Chem. Eng. J.* **2015**, *277*, 86–96.

12. Che, H.; Lee, W. Selective redox degradation of chlorinated aliphatic compounds by Fenton reaction in pyrite suspension. *Chemosphere* **2011**, *82*, 1103–1108.

13. Huang, R.; Fang, Z.; Yan, X.; Cheng, W. Heterogeneous sono-Fenton catalytic degradation of bisphenol A by Fe3O4 magnetic nanoparticles under neutral condition. *Chem. Eng. J.* **2012**, *197*, 242–249.

14. Masomboon, N.; Ratanatamskul, C.; Lu, M.C. Chemical oxidation of 2,6-dimethylaniline in the Fenton process. *Environ. Sci. Technol.* **2009**, *43*, 8629–8634.

15. Matta, R.; Hanna, K.; Chiron, S. Fenton-like oxidation of 2,4,6-trinitrotoluene using different iron minerals. *Sci. Total. Environ.* **2007**, *385*, 242–251.

16. Maria, A.F.D.; Emilio, R.; María, F.F.; Marta, P.; Maria, Á.S. Degradation of organic pollutants by heterogeneous electro-Fenton process using Mn-alginate composite. *J. Chem. Technol. Biotechnol.*

17. Hu, X.B.; Deng, Y.H.; Gao, Z.Q.; Liu, B.Z.; Sun, C. Transformation and reduction of androgenic activity of 17 alpha-methyltestosterone in Fe3O4/MWCNTs-H2O2 system. *Appl. Catal. B-Environ.* **2012**, *127*, 167–174.

18. Todd, E. Surface oxidation of pyrite under ambient atmo-spheric and aqueous (pH = 2 to 10) conditions: Electronic structure and mineralogy from X-ray absorption spectroscopy. *Geochim. Cosmochim. Acta* **2003**, *67*, 881–893.

19. Arienzo, M. Oxidizing 2,4,6-trinitroluene with pyrite-H2O2 suspensions. *Chemosphere* **1999**, *39*, 1629–1638.

20. Che, H.; Bae, S.; Lee, W. Degradation of trichloroethylene by Fenton reaction in pyrite suspension. *J. Hazard. Mater.* **2011**, *185*, 1355–1361.

21. Bae, S.; Kim, D.; Lee, W. Degradation of diclofenac by pyrite catalyzed Fenton oxidation. *Appl. Catal. B-Environ.* **2013**, *134*, 93–102.

22. Wu, D.L.; Feng, Y.; Ma, L.M. Oxidation of Azo Dyes by H2O2 in presence of natural pyrite. *Water Air Soil. Pollut.* **2013**, *224*, doi:10.1007/s11270-012-1407-y.

23. Zhang, Y.L.; Zhang, K.; Dai, C.M.; Zhou, X.F.; Si, H.P. An enhanced Fenton reaction catalyzed by natural heterogeneous pyrite for nitrobenzene degradation in an aqueous solution. *Chem. Eng. J.* **2014**, *244*, 438–445.

24. Wang, C.; Xi, J.Y.; Hu, H.Y. Chemical identification and acute biotoxicity assessment of gaseous chlorobenzene photodegradation products. *Chemosphere* **2008**, *73*, 1167–1171.

25. APHA; AWWA; WEF. *Standard Methods for the Examination of Water and Wastewater*, 20th ed.; APHA/AWWA/WEF: Washington, DC, USA, 1998.

26. Wang, L.S.; Wei, D.B.; Wei, J.; Hu, H.Y. Screening and estimating of toxicity formation with photobacterium bioassay during chlorine disinfection of wastewater. *J. Hazard. Mater.* **2007**, *141*, 289–294.

27. Stookey, L.L. Ferrozine — A new spectrophotometric reagent for iron. *Anal. Chem.* **1970**, *42*, 779–781.

28. Norwitz, G.; Keliher, P.N. Spectrophotometric determination of aniline by the diazotization-coupling method with N-(1-naphthyl) ethylenediamine as the coupling agent. *Anal. Chem.* **1981**, *53*, 1238–1240.

29. Stedmon, C.A.; Bro, R. Characterizing dissolved organic matter fluorescence with parallel factor analysis: A tutorial. *Limnol. Oceanogr. Methods* **2008**, *6*, 572–579.

30. Baghoth, S.A.; Sharma, S.K.; Amy, G.L. Tracking natural organic matter (NOM) in a drinking water treatment plant using fluorescence excitation — Emission matrices and PARAFAC. *Water Res.* **2011**, *45*, 797–809.

31. Li, W.T.; Chen, S.Y.; Xu, Z.X.; Li, Y.; Shuang, C.D.; Li, A.M. Characterization of dissolved organic matter in municipal wastewater using fluorescence PARAFAC analysis and chromatography multi-excitation/emission scan: A comparative study. *Environ. Sci. Technol.* **2014**, *48*, 2603–2609.

32. Pignatello, J.J.; Oliveros, E.; MacKay, A. Advanced oxidation processes for organic contaminant destruction based on the Fenton reaction and related chemistry. *Crit. Rev. Environ. Sci. Technol.* **2006**, *36*, 1–84.

33. Hashemian, S.; Tabatabaee, M.; Gafari, M. Fenton oxidation of methyl violet in aqueous solution. *J. Chem.* **2013**, *2013*, doi:10.1155/2013/509097.

34. Zhang, Y.L.; Zhang K.; Dai, C.M.; Zhou, X.F. Performance and mechanism of pyrite for nitrobenzene removal in aqueous solution. *Chem. Eng. Sci.* **2014**, *111*, 135–141.

35. Segura, Y.; Martínez, F.; Melero, J.A.; Fierro, J.L.G. Zero valent iron (ZVI) mediated Fenton degradation of industrial wastewater: Treatment performance and characterization of final composites. *Chem. Eng. J.* **2015**, *269*, 298–305.

36. Moffett, J.W.; Zika, R.G. Reaction kinetics of hydrogen peroxide with copper and iron in seawater. *Environ. Sci. Technol.* **1987**, *21*, 804–810.

37. Stumm, W.; James, J. *Aquatic Chemistry. An Introduction Empha – Sizing Chemical Equilibria in Natural Waters*; John Wiley & Sons: Michigan, MI, USA, 1981.

38. Pham, A.L.T.; Lee, C.; Doyle, F.M.; Sedlak, D.L. A silica-supported iron oxide catalyst capable of activating hydrogen peroxide at neutral pH values. *Environ. Sci. Technol.* **2009**, *43*, 8930–8935.

39. Xue, X.X.; Hanna, K.; Abdelmoula, M.; Deng, N. Adsorption and oxidation of PCP on the surface of magnetite: Kinetic experiments and spectroscopic investigations. *Appl. Catal. B-Environ.* **2009**, *89*, 432–440.

40. Kong, L.H.; Hu, X.Y.; He, M.C. Mechanisms of Sb(III) oxidation by pyrite-induced hydroxyl radicals and hydrogen peroxide. *Environ. Sci. Technol.* **2015**, *49*, 3499–3505.

41. Kwan, W.P.; Voelker, B.M. Decomposition of hydrogen peroxide and organic compounds in the presence of dissolved iron and ferrihydrite. *Environ. Sci. Technol.* **2002**, *36*, 1467–1476.

42. Westerhoff, P.; Chen, W.; Esparza, M. Fluorescence analysis of a standard fulvic acid and tertiary treated wastewater. *J. Environ. Qual.* **2001**, *30*, 2037–2046.

43. Goldman, J.H.; Rounds, S.A.; Needoba, J.A. Applications of fluorescence spectroscopy for predicting percent wastewater in an urban stream. *Environ. Sci. Technol.* **2012**, *46*, 4374–4381.

44. Stedmon, C.A.; Markager, S.; Bro, R. Tracing dissolved organic matter in aquatic environments using a new approach to fluorescence spectroscopy. *Mar. Chem.* **2003**, *82*, 239–254.

45. Bro, R. The N-Way On-Line Course on PARAFAC and PLS. Available online: http://www.models.life.ku.dk/~pih/parafac/chap0contents.htm (accessed on 21 June 2010).

46. Lei, Y.Q.; Wang, G.H.; Guo, P.R.; Li, G.B.; Cai, D.C. The application of three-dimensional fluorescence spectroscopy in the electrochemical degradation of organic pollutions. *Chin. Jled. Anal. Lab.* **2014**, *33*, 373–376. (In Chinese).

This page left intentionally blank.

INDEX